D2w   12/02   £52.50

FLUID POWER DYNAMICS

FLUID POWER DYNAMICS

R. Keith Mobley

Newnes

Boston   Oxford   Auckland   Johannesburg   Melbourne   New Delhi

Library of Congress Cataloging-in-Publication Data

Mobley, R. Keith, 1943-
    Fluid power dynamics / by R. Keith Mobley.
       p. cm. — (Plant engineering maintenance series)
    Includes index.
    ISBN 0-7506-7174-2 (alk. paper)
    1. Fluid power technology.   I. Title.   II. Series.
  TJ843.M63 1999
  620.1'06—dc21   98-35357
                                     CIP

British Library Cataloguing-in-Publication Data
A catalogue record for this book is available from the British Library.

The publisher offers special discounts on bulk orders of this book.
For information, please contact:
Manager of Special Sales
Butterworth–Heinemann
225 Wildwood Avenue
Woburn, MA 01801–2041
Tel: 781-904-2500
Fax: 781-904-2620

For information on all Newnes publications available, contact our World Wide Web home page at: http://www.newnespress.com

10 9 8 7 6 5 4 3 2 1

Printed in the United States of America

CONTENTS

INTRODUCTION

The study of hydraulics deals with the use and characteristics of liquids and gases. Since the beginning of time, humans have used fluids to ease their burdens. Earliest recorded history shows that devices such as pumps and waterwheels were used to generate useable mechanical power.

Fluid power encompasses most applications that use liquids or gases to transmit power in the form of mechanical work, pressure, and/or volume in a system. This definition includes all systems that rely on pumps or compressors to transmit specific volumes and pressures of liquids or gases within a closed system. The complexity of these systems ranges from a simple centrifugal pump used to remove casual water from a basement to complex airplane control systems that rely on high-pressure hydraulic systems.

Fluid power systems have been developing rapidly over the past 35 years. Fluid power filled a need during World War II for an energy transmission system with muscle, which could easily be adapted to automated machinery. Today, fluid power technology is seen in every phase of human activity. Fluid power is found in areas of manufacturing such as metal forming, plastics, basic metals, and material handling. Fluid power is evident in transportation as power and control systems of ships, airplanes, and automobiles. The environment is another place fluid power is hard at work compacting waste materials and controlling floodgates of hydroelectric dams. Food processing, construction equipment, and medical technology are a few more areas of fluid power involvement. Fluid power applications are only limited by imagination.

There are alternatives to fluid power systems. Each system, regardless of the type, has its own advantages and disadvantages. Each has applications where it is best suited to do the job. This is probably the reason you won't find a fluid power wristwatch, or hoses carrying fluid power replacing electrical power lines.

ADVANTAGES OF FLUID POWER

If a fluid power system is properly designed and used, it will provide smooth, flexible, uniform action without vibration and is unaffected by variation of load. In case of an overload, an automatic release of pressure can be guaranteed, so that the system is protected against breakdown or excessive strain. Fluid power systems can provide widely variable motions in both rotary and linear transmission of power, and the need for manual control can be minimized. In addition, fluid power systems are economical to operate.

Fluid power includes hydraulic, hydro-pneumatic, and pneumatic systems. Why are hydraulics used in some applications, pneumatics in others, or combination systems in still others? Both the user and the manufacturer must consider many factors when determining which type of system should be used in a specific application.

In general, pneumatic systems are less expensive to manufacture and operate, but there are factors that prohibit their universal application. The compressibility of air, like that of any gas, limits the operation of pneumatic systems. For example, a pneumatic cylinder cannot maintain the position of a suspended load without a constant supply of air pressure. The load will force the air trapped within the cylinder to compress and allow the suspended load to creep. This compressibility also limits the motion of pneumatic actuators when under load.

Pneumatic systems can be used for applications that require low to medium pressure and only fairly accurate control. Applications that require medium pressure, more accurate force transmission, and moderate motion control can use a combination of hydraulics and pneumatics, or hydro-pneumatics. Hydraulics systems must be used for applications that require high pressure and/or extremely accurate force and motion control.

The flexibility of fluid power, both hydraulic and pneumatic, elements presents a number of problems. Since fluids and gases have no shape of their own, they must be positively confined throughout the entire system. This is especially true in hydraulics, where leakage of hydraulic oil can result in safety or environmental concerns. Special consideration must be given to the structural integrity of the parts of a hydraulic system. Strong pipes, tubing, and hoses, as well as strong containers, must be provided. Leaks must be prevented. This is a serious problem with the high pressure obtained in many hydraulic system applications.

Fluid Power Systems vs Mechanical Systems

Fluid power systems have some desirable characteristics when compared with mechanical systems:

A fluid power system is often a simpler means of transmitting energy. There are fewer mechanical parts in an ordinary industrial system. Since there are fewer mechanical parts, a fluid power system is more efficient and more dependable. In the common industrial system, there is no need to worry about hundreds of moving parts failing, with fluid or gas as the transmission medium.

With fluid or gas as the transmission medium, various components of a system can be located at convenient places on the machine. Fluid power can be transmitted and controlled quickly and efficiently up, down, and around corners with few controlling elements.

Since fluid power is efficiently transmitted and controlled, it gives freedom in designing a machine. The need for gear, cam, and lever systems is eliminated. Fluid power systems can provide infinitely variable speed, force and direction control with simple, reliable elements.

Fluid Power vs Electrical Systems

Mechanical force and motion controlled can be more easily controlled using fluid power. The simple use of valves and rotary or linear actuators controls speed, direction, and force. The simplicity of hydraulic and pneumatic components greatly increases their reliability. In addition, smaller components and overall system size are typically much smaller than comparable electrical transmission devices.

SPECIAL PROBLEMS

The operation of the system involves constant movement of the hydraulic fluid within its lines and components. This movement causes friction within the fluid itself and against the containing surfaces. Excessive friction can lead to serious losses in efficiency or damage to system components. Foreign matter must not be allowed to accumulate in the system, where it will clog small passages or score closely fitted parts. Chemical action may cause corrosion. Anyone working with hydraulic systems must know how a fluid power system and its components operate, both in terms of the general principles common to all physical mechanisms and in terms of the peculiarities of the specific arrangement at hand.

The word hydraulics is based on the Greek word for water, the first-used form of hydraulic power transmission. Initially, hydraulics covered the study of the physical behavior of water at rest and in motion. It has been expanded to include the behavior of all liquids, although it is primarily limited to the motion or kinetics of liquids.

HAZARDS

Any use of a pressurized medium, such as hydraulic fluid, can be dangerous. Hydraulic systems carry all the hazards of pressurized systems and special hazards related directly to the composition of the fluid used.

When oil is used as a fluid in a high-pressure hydraulic system, the possibility of fire or an explosion exists. A severe fire hazard is generated when a break in the high-pressure piping occurs and the oil is vaporized into the atmosphere. Extra precautions against fire should be practiced in these areas.

If oil is pressurized by compressed air, an explosive hazard exists. If high-pressure air comes into contact with the oil, it may create a diesel effect, which may result in an explosion. A carefully followed preventive maintenance plan is the best precaution against explosions.

Part I

HYDRAULICS

1

BASIC HYDRAULICS

Fluid power systems have developed rapidly over the past 35 years. Today, fluid power technology is used in every phase of human existence. The extensive use of hydraulics to transmit power is due to the fact that properly constructed fluid power systems possess a number of favorable characteristics. They eliminate the need for complicated systems of gears, cams, and levers. Motion can be transmitted without the slack or mechanical looseness inherent in the use of solid machine parts. The fluids used are not subject to breakage as are mechanical parts, and the mechanisms are not subjected to great wear.

The operation of a typical fluid power system is illustrated in Figure 1-1. Oil from a tank or reservoir flows through a pipe into a pump. An electric motor, air motor, gas or steam turbine, or an internal combustion engine can drive the pump. The pump increases the pressure of the oil. The actual pressure developed depends on the design of the system.

The high-pressure oil flows in piping through a control valve. The control valve changes the direction of oil flow. A relief valve, set at a desired, safe operating pressure, protects the system from an overpressure condition. The oil that enters the cylinder acts on the piston, with the pressure acting over the area of the piston, developing a force on the piston rod. The force on the piston rod enables the movement of a load or device.

STATES OF MATTER

The material that makes up the universe is known as matter. Matter is defined as any substance that occupies space and has weight. Matter exists in three states: solid, liquid, and gas. Each has distinguishing characteristics. Solids have a defined volume and a definite shape. Liquids have a definite volume, but take the shape of their containing vessels. Gases have neither a definite shape nor a definite volume. Gases not

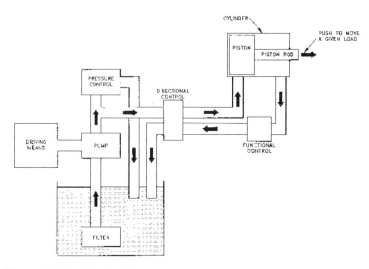

Figure 1–1 Basic hydraulic system.

only take the shape of the containing vessel, but also expand to fill the vessel, regardless of its volume. Examples of the states of matter are iron, water, and air.

Matter can change from one state to another. Water is a good example. At high temperatures, above 212° Fahrenheit (°F), it is in a gaseous state known as steam. At moderate temperatures, it is liquid, and at low temperatures, below 32°F, it becomes ice, a solid. In this example, the temperature is the dominant factor in determining the state that the substance assumes.

Pressure is another important factor that will affect changes in the state of matter. At pressures lower than atmospheric, 14.7 psi, water will boil and thus change to steam at temperatures below 212°F. Pressure is also a critical factor in changing some gases to liquids or solids. Normally, when pressure and chilling are both applied to a gas, the gas assumes a liquid state. Liquid air, which is a mixture of oxygen and nitrogen, is produced in this manner.

In the study of fluid power, we are concerned primarily with the properties and characteristics of liquids and gases. However, you should keep in mind that the properties of solids also affect the characteristics of liquids and gases. The lines and components, which are solids, enclose and control the liquid or gas in their respective systems.

DEVELOPMENT OF HYDRAULICS

The use of hydraulics is not new. The Egyptians and people of ancient Persia, India, and China conveyed water along channels for irrigation and other domestic purposes. They used dams and sluice gates to control the flow and waterways to direct the water to where it was needed. The ancient Cretans had elaborate plumbing systems. Archimedes studied the laws of floating and submerged bodies. The Romans constructed aqueducts to carry water to their cities.

After the breakup of the ancient world, there were few new developments for many centuries. Then, over a comparatively short period, beginning near the end of the seventeenth century, Italian physicist Evangelista Torricelli, French physicist Edme Mariotte, and later Daniel Bernoulli conducted experiments to study the force generated by the discharge of water through small openings in the sides of tanks and through short pipes. During the same period, Blaise Pascal, a French scientist, discovered the fundamental law for the science of hydraulics. Pascal's law states that an increase in pressure on the surface of a confined fluid is transmitted throughout the confining vessel or system without any loss of pressure.

Figure 1-2 illustrates the transmission of forces through liquids. For Pascal's law to become effective for practical applications, a piston or ram confined within a close tolerance cylinder was needed. It was not until the latter part of the eighteenth century that methods were developed that could manufacture the snugly fitted parts required to make hydraulic systems practical.

This was accomplished by the invention of machines that were used to cut and shape the necessary closely fitted parts, and particularly by the development of gaskets and packing. Since that time, components such as valves, pumps, actuating cylinders, and motors have been developed and refined to make hydraulics one of the leading methods of transmitting power.

USE OF HYDRAULICS

The hydraulic press, invented by Englishman John Brahmah, was one of the first workable machines that used hydraulics in its operation. It consisted of a plunger pump piped to a large cylinder and a ram. This press found wide use in England because it provided a more effective and economical means of applying large, uniform forces in industrial uses.

Today, hydraulic power is used to operate many different tools and mechanisms. In a garage, a mechanic raises the end of an automobile with a hydraulic jack. Dentists and

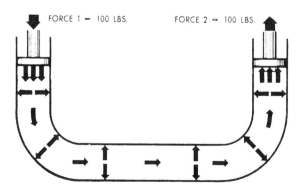

Figure 1–2 Transmission of forces.

barbers use hydraulic power to lift and position their chairs. Hydraulic doorstops keep heavy doors from slamming. Hydraulic brakes have been standard equipment on automobiles since the 1930s. Most automobiles are equipped with automatic transmissions that are hydraulically operated. Power steering is another application of hydraulic power. Construction workers depend upon hydraulic power for their equipment. For example, the blade of a bulldozer is normally operated by hydraulic power.

Operation of Hydraulic Components

To transmit and control power through pressurized fluids, an arrangement of interconnected components is required. Such an arrangement is commonly referred to as a system. The number and arrangement of the components vary from system to system, depending on the particular application. In many applications, one main system supplies power to several subsystems, which are sometimes referred to as circuits. The complete system may be a small, compact unit or a large, complex system that has components located at widely separated points within the plant. The basic components of a hydraulic system are essentially the same, regardless of its complexity. These seven components, which must be in every hydraulic system, are as follows:

> *Reservoir or receiver.* This is usually a closed tank or vessel that holds the volume of fluid required to support the system. The vessels normally provide several functions in addition to holding fluid reserves. The major functions include filtration of the fluid, heat dissipation, and water separation.

> *Hydraulic pump.* This is the energy source for hydraulic systems. It converts electrical energy into dynamic, hydraulic pressure. In almost all cases, hydraulic systems utilize positive displacement pumps as their primary power source. These are broken down into two primary subclassifications: constant-volume or variable-volume. In the former, the pumps are designed to deliver a fixed output (i.e., both volume and pressure) of hydraulic fluid. In the latter, the pump delivers only the volume or pressure required for specific functions of the system or its components.

> *Control valves.* The energy generated by the hydraulic pump must be directed and controlled so that the energy can be used. There are a variety of directional and functional control valves that are designed to provide a wide range of control functions.

> *Actuating devices.* The energy within a hydraulic system is of no value until it is converted into work. Typically, this is accomplished by using an actuating device of some type. This actuating device may be a cylinder, which converts the hydraulic energy into linear mechanical force; a hydraulic motor, which converts energy into rotational force; or a variety of other actuators designed to provide specific work functions.

> *Relief valves.* Most hydraulic systems use a positive displacement pump to generate energy within the system. Unless the pressure is controlled, these

pumps will generate excessive pressure that can cause catastrophic failure of system components. A relief valve is always installed downstream of the hydraulic pump to prevent excessive pressure and to provide a positive relief should a problem develop within the system. The relief valve is designed to open at a preset system pressure. When the valve opens, it diverts flow to the receiver tank or reservoir.

Lines (pipe, tubing, or flexible hoses). All systems require some means to transmit hydraulic fluid from one component to another. The material of the connecting lines will vary from system to system or within the system.

Hydraulic fluid. The fluid provides the vehicle that transmits input power, such as from a hydraulic pump to the actuator device or devices that perform work.

2

FORCES IN LIQUIDS

The study of liquids is divided into two main parts: liquids at rest, hydrostatics; and liquids in motion, hydraulics. The effect of liquids at rest can often be expressed by simple formulas. The effects of liquids in motion are more difficult to express because of frictional and other factors whose actions cannot be expressed by simple mathematics.

Liquids are almost incompressible. For example, if a pressure of 100 pounds per square inch, psi, is applied to a given volume of water that is at atmospheric pressure, the volume will decrease by only 0.03 percent. It would take a force of approximately 32 tons to reduce its volume by 10 percent; however, when this force is removed, the water immediately returns to its original volume. Other liquids behave in about the same manner as water.

Another characteristic of a liquid is the tendency to keep its free surface level. If the surface is not level, liquids will flow in the direction that will tend to make the surface level.

LIQUIDS AT REST (HYDROSTATICS)

In the study of fluids at rest, we are concerned with the transmission of force and the factors that affect the forces in liquids. Additionally, pressure in and on liquids and factors affecting pressure are of great importance.

PRESSURE AND FORCE

The terms force and pressure are used extensively in the study of fluid power. It is essential that we distinguish between these terms. Force is the total pressure applied to or generated by a system. It is the total pressure exerted against the total area of a particular surface and is expressed in pounds or grams.

Pressure is the amount of force applied to each unit area of a surface and is expressed in pounds per square inch, lb/in^2 (psi) or grams per square centimeter, gm/cm^2. Pressure may be exerted in one direction, in several directions, or in all directions.

A formula is used in computing force, pressure, and area in fluid power systems. In this formula, P refers to pressure, F indicates force, and A represents area. Force equals pressure times area. Thus, the formula is written

$$F = P \times A$$

Pressure equals force divided by area. By rearranging the formula, this statement may be condensed to

$$P = \frac{F}{A}$$

Since area equals force divided by pressure, the formula is written

$$A = \frac{F}{P}$$

ATMOSPHERIC PRESSURE

The atmosphere is the entire mass of air that surrounds the earth. Although it extends upward for about 500 miles, the section of primary interest is the portion that rests on the earth's surface and extends upward for about 7 1/2 miles. This layer is called the troposphere.

If a column of air 1 inch square extended to the "top" of the atmosphere could be weighed, this column of air would weigh approximately 14.7 pounds at sea level. Thus, atmospheric pressure, at sea level, is approximately 14.7 pounds per square inch or psi.

Atmospheric pressure decreases by approximately 1.0 psi for every 2,343 feet of elevation. At elevations below sea level, such as in excavations and depressions, atmospheric pressure increases. Pressures under water differ from those under air only because the weight of the water must be added to the pressure of the air.

Atmospheric pressure can be measured by any of several methods. The common laboratory method uses a mercury column barometer. The height of the mercury column serves as an indicator of atmospheric pressure. At sea level and at a temperature of 0° Celsius (°C), the height of the mercury column is approximately 30 inches, or 76 centimeters. This represents a pressure of approximately 14.7 psia. The 30-inch column is used as a reference standard.

Atmospheric pressure does not vary uniformly with altitude. It changes more rapidly at lower altitudes because of the compressibility of air, which causes the air layers close to the earth's surface to be compressed by the air masses above them. This effect, however, is partially counteracted by the contraction of the upper layers due to cooling. The cooling tends to increase the density of the air.

Atmospheric pressures are quite large, but in most instances practically the same pressure is present on all sides of objects so that no single surface is subjected to a greater load. Atmospheric pressure acting on the surface of a liquid (Figure 2-1A) is transmitted equally throughout the liquid to the walls of the container, but is balanced by the same atmospheric pressure acting on the outer walls of the container. In part B of Figure 2-1, atmospheric pressure acting on the surface of one piston is balanced by the same pressure acting on the surface of the other piston. The different areas of the two surfaces make no difference, since for a unit of area, pressures are balanced.

Pascal's Law

The foundation of modern hydraulics was established when Pascal discovered that pressure in a fluid acts equally in all directions. This pressure acts at right angles to the containing surfaces. If some type of pressure gauge, with an exposed face, is placed beneath the surface of a liquid (Figure 2-2) at a specific depth and pointed in different directions, the pressure will read the same. Thus, we can say that pressure in a liquid is independent of direction.

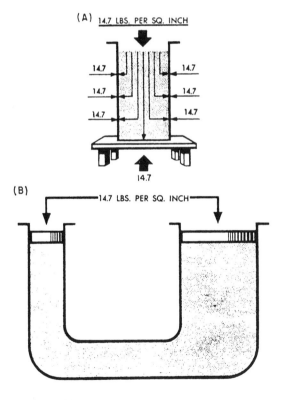

Figure 2–1 Effects of atmospheric pressure.

Pressure due to weight of a liquid, at any level, depends on the depth of the fluid from the surface. If the exposed face of the pressure gauge in Figure 2-2 is moved closer to the surface of the liquid, the indicated pressure will be less.

When the depth is doubled, the indicated pressure is also doubled. Thus, the pressure in a liquid is directly proportional to the depth. Consider a container with vertical sides (Figure 2-3) that are 1 foot high and 1 foot wide.

Let it be filled with water 1 foot deep, thus providing 1 cubic foot of water. We learned earlier in this chapter that 1 cubic foot of water weighs 62.4 pounds. Using

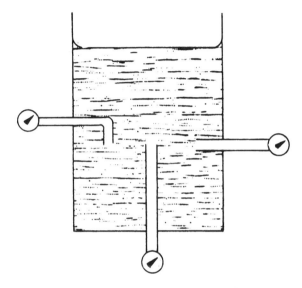

Figure 2–2 Pressure of a liquid is independent of direction.

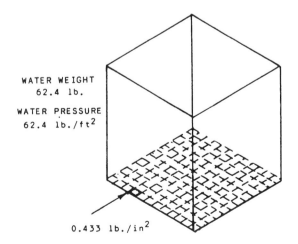

WATER WEIGHT
62.4 lb.

WATER PRESSURE
62.4 lb./ft^2

0.433 lb./in^2

Figure 2–3 Water pressure in a 1-cubic-foot container.

this information and the equation for pressure we can calculate the pressure on the bottom of the container.

$$P = \frac{F}{A} = \frac{62.4 \text{ lbs.}}{1 \text{ ft}^2}$$

$$P = \frac{62.4}{144} = 0.433 \text{ lbs/in}^2$$

Since there are 144 square inches in 1 square foot, this can be stated as follows: The weight of a column of water 1 foot high, having a cross-sectional area of 1 square inch, is 0.433 pounds. If the depth of the column is tripled, the weight of the column will be 3×0.433 or 1.299 pounds and the pressure at the bottom will be 1.299 lb/in^2 (psi), since the pressure is equal to the force divided by the area. Thus, the pressure at any depth in a liquid is equal to the weight of the column of liquid at the depth divided by the cross-sectional area of the column at that depth. The volume of a liquid that produces the pressure is referred to as the fluid head of the liquid. The pressure of a liquid due to its fluid head is also dependent on the density of the liquid.

If we let A equal any cross-sectional area of a liquid column and h equal the depth of the column, the volume becomes Ah. Using the equation $D = W/V$, the weight of the liquid above area A is equal to AhD, or

$$D = \frac{W}{Ah} W = Ah \times D$$

Since pressure is equal to the force per unit area, set A equal to 1. Then the formula for pressure becomes

$$P = hD$$

It is essential that h and D be expressed in similar units. That is, if D is expressed in pounds per cubic foot, the value of h must be expressed in feet. If the desired pressure is to be expressed in pounds per square inch, the pressure formula becomes

$$P = \frac{hD}{144}$$

Pascal was also the first to prove by experiment that the shape and volume of a container in no way alters pressure. Thus, in Figure 2-4, if the pressure due to the weight of the liquid at a point on horizontal line H is 8 psi, the pressure is 8 psi everywhere at level H in the system.

LIQUIDS IN MOTION (HYDRAULICS)

In the operation of fluid power systems, there must be flow of fluid. The amount of flow will vary from system to system. To understand fluid power systems, it is necessary to understand some of the characteristics of liquids in motion.

Liquids in motion have characteristics different from those of liquids at rest. Frictional resistances within the fluid, viscosity, and inertia contribute to these differences. Inertia, which

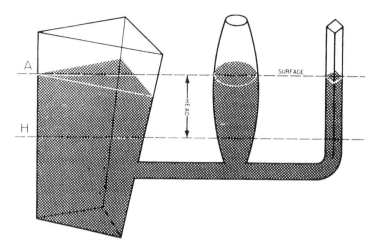

Figure 2–4 Pressure relationship with shape.

means the resistance a mass offers to being set in motion, will be discussed later in this section. There are other relationships of liquids in motion with which you must become familiar. Among these are volume and velocity of flow; flow rate and speed; laminar and turbulent flow; and more importantly, the force and energy changes that occur in flow.

VOLUME AND VELOCITY OF FLOW

The volume of a liquid passing a point in a given time is known as its volume of flow or flow rate. The volume of flow is usually expressed in gallons per minute (gpm) and is associated with the relative pressures of the liquid, such as 5 gpm at 40 psig. The velocity of flow or velocity of the fluid is defined as the average speed at which the fluid moves past a given point. It is usually expressed in feet per minute (fpm) or inches per second (ips). Velocity of flow is an important consideration in sizing the hydraulic piping and other system components.

Volume and velocity of flow must be considered together. With other conditions unaltered, the velocity of flow increases as the cross-section or size of the pipe decreases, and the velocity of flow decreases as the cross-section or pipe size increases. For example, the velocity of flow is slow at wide parts, yet the volume of liquid passing each part of the stream is the same.

In Figure 2-5, if the cross-sectional area of the pipe is 16 square inches at point A and 4 square inches at point B, we can calculate the relative velocity of flow using the flow equation

$$Q = vA$$

where Q is the volume of flow, v is the velocity of flow, and A is the cross-sectional area of the liquid. Since the volume of flow at point A, Q_1, is equal to the volume of

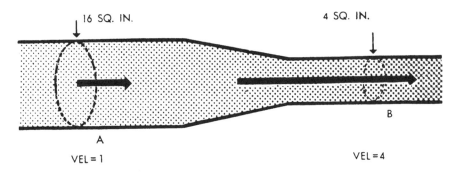

Figure 2–5 Volume and velocity of flow.

flow at point B, Q_2, we can use this equation to determine the ratio of the velocity of flow at point A, v_1, to the velocity of flow at point B, v_2.

Since, $Q_1 = Q_2$, then $A_1v_1 = A_2v_2$

From Figure 2-5:

$A_1 = 16$ square inches, $A_2 = 4$ square inches

Substituting: $16v_1 = 4v_2$ or $v_2 = 4v_1$

Therefore, the velocity of flow at point B is four times greater than the velocity of flow at point A.

VOLUME OF FLOW AND SPEED

When you consider the cylinder volume that must be filled and the distance that the piston must travel, you can relate the volume of flow to the speed of the piston. The volume of the cylinder is found by multiplying the piston area by the length the piston must travel. This length is known as stroke.

Suppose you have determined that two cylinders have the same volume and that one cylinder is twice as long as the other. In this case, the cross-sectional area of the longer tube will be half of the cross-sectional area of the other tube. If fluid is pumped into each cylinder at the same rate, both pistons will reach their full travel at the same time. However, the piston in the smaller cylinder must travel twice as fast because it has twice as far to go. There are two ways of controlling the speed of the piston: (1) by varying the size of the cylinder and (2) by varying the volume of flow (gpm) to the cylinders.

STREAMLINE AND TURBULENT FLOW

At low velocities or in tubes of small diameter, flow is streamlined. This means that a given particle of fluid moves straight forward without bumping into other particles and without crossing their paths. Streamline flow is often referred to as laminar flow, which is defined as a flow situation in which fluid moves in parallel lamina or layers. As an example of streamline flow, consider an open stream flowing at a slow, uniform rate with logs floating on its surface. The logs represent particles of fluid. As long as the stream flows at a slow, uniform rate, each log floats downstream in its own path, without crossing or bumping into the others.

If the stream narrows and the volume of flow remains the same, the velocity will increase. If the velocity increases sufficiently, the water becomes turbulent. Swirls, eddies, and cross-motions are set up in the water. As this happens, the logs are thrown against each other and against the banks of the stream, and the paths followed by different logs will cross and recross.

Particles of fluid flowing in pipes act in the same manner. The flow is streamline if the fluid flows slowly enough, and remains streamline at greater velocities if the diameter of the pipe is small. If the velocity of flow or size of pipe is increased sufficiently, the flow becomes turbulent.

Although a high velocity of flow will produce turbulence in any pipe, other factors contribute to turbulence. Among these are the roughness of the inside of the pipe, obstructions, and the number and degree of curvature of bends in the pipe. In setting up or maintaining fluid power systems, care should be taken to eliminate or minimize as many causes of turbulence as possible, since energy consumed by turbulence is wasted.

Although designers of fluid power equipment do what they can to minimize turbulence, it cannot be avoided. For example, in a 4-inch pipe at 680°F, flow becomes turbulent at velocities over approximately 6 inches per second (ips) or about 3 ips in a 6-inch pipe. These velocities are far below those commonly encountered in fluid power systems, where velocities of 5 feet per second (fps) and above are common. In laminar flow, losses due to friction increase directly with velocity. With turbulent flow, these losses increase much more rapidly.

FACTORS INVOLVED IN FLOW

An understanding of the behavior of fluids in motion, or solids for that matter, requires an understanding of the term inertia. Inertia is the term used by scientists to describe the property possessed by all forms of matter that make it resist being moved when it is at rest and resist any change in its rate of motion when it is moving.

The basic statement covering inertia is Newton's first law of motion. His first law states: A body at rest tends to remain at rest, and a body in motion tends to remain in motion at the same speed and direction, unless acted on by some unbalanced force. This simply says what you have learned by experience—that you must push an object to start it moving and push it in the opposite direction to stop it again.

A familiar illustration is the effort a pitcher must exert to make a fast pitch and the opposition the catcher must put forth to stop the ball. Similarly, the engine to make an automobile begin to roll must perform considerable work—although, after it has attained a certain velocity, it will roll along the road at uniform speed if just enough effort is expended to overcome friction, while brakes are necessary to stop its motion. Inertia also explains the kick or recoil of guns and the tremendous striking force of projectiles.

INERTIA AND FORCE

To overcome the tendency of an object to resist any change in its state of rest or motion, some force that is not otherwise canceled or balanced must act on the object. Some unbalanced force must be applied whenever fluids are set in motion or increased in velocity; conversely, forces are made to do work elsewhere whenever fluids in motion are retarded or stopped.

There is a direct relationship between the magnitude of the force exerted and the inertia against which it acts. This force is dependent on two factors: (1) the mass of the object and (2) the rate at which the velocity of the object is changed. The rule is that the force, in pounds, required to overcome inertia is equal to the weight of the object multiplied by the change in velocity, measured in feet per second (fps) and divided by 32 times the time, in seconds, required to accomplish the change. Thus, the rate of change in velocity of an object is proportional to the force applied. The number 32 appears because it is the conversion factor between weight and mass.

There are five physical factors that can act on a fluid to affect its behavior. All of the physical actions of fluids in all systems are determined by the relationship of these five factors to each other:

1. Gravity, which acts at all times on all bodies, regardless of other forces
2. Atmospheric pressure, which acts on any part of a system exposed to the open air
3. Specific applied forces, which may or may not be present, but which are entirely independent of the presence or absence of motion
4. Inertia, which comes into play whenever there is a change from rest to motion, or the opposite, or whenever there is a change in direction or in rate of motion
5. Friction, which is always present whenever there is motion

KINETIC ENERGY

An external force must be applied to an object in order to give it a velocity or to increase the velocity it already has. Whether the force begins or changes velocity, it acts over a certain distance. Force acting over a certain distance is called work. Work and all forms into which it can be changed are classified as energy. Obviously, then, energy is required to give an object velocity. The greater the energy used, the greater the velocity will be.

Disregarding friction, for an object to be brought to rest or for its motion to be slowed down, a force opposed to its motion must be applied to it. This force also acts over some distance. In this way energy is given up by the object and delivered in some form to whatever opposed its continuous motion. The moving object is therefore a means of receiving energy at one place and delivering it to another point. While it is in motion, it is said to contain this energy, as energy of motion or kinetic energy.

Since energy can never be destroyed, it follows that if friction is disregarded, the energy delivered to stop the object will exactly equal the energy that was required to increase its speed. At all times, the amount of kinetic energy possessed by an object depends on its weight and the velocity at which it is moving.

The mathematical relationship for kinetic energy is stated in the following rule: Kinetic energy, in foot-pounds, is equal to the force, in pounds, that created it, multiplied by the distance through which it was applied; or it is equal to the weight of the moving object, in pounds, multiplied by the square of its velocity in feet per second, and divided by 64.

The relationship between inertia forces, velocity, and kinetic energy can be illustrated by analyzing what happens when a gun fires a projectile against the armor of an enemy ship. The explosive force of the powder in the breech pushes the projectile out of the gun, giving it a high velocity. Because of its inertia, the projectile offers opposition to this sudden velocity and a reaction is set up that pushes the gun backwards. The force of the explosion acts on the projectile throughout its movement in the gun. This is force acting through a distance, producing work. This work appears as kinetic energy in the speeding projectile. The resistance of the air produces friction, which uses some of the energy and slows down the projectile. When the projectile hits the target, it tries to continue moving. The target, being relatively stationary, tends to remain stationary because of inertia. The result is that a tremendous force is set up that leads either to the penetration of the armor or to the shattering of the projectile. The projectile is simply a means to transfer energy from the gun to the enemy ship. This energy is transmitted in the form of energy in motion or kinetic energy.

A similar action takes place in a fluid power system in which the fluid takes the place of the projectile. For example, the pump in a hydraulic system imparts energy to the fluid, which overcomes the inertia of the fluid at rest and causes it to flow through the lines. The fluid flows against some type of actuator that is at rest. The fluid tends to continue flowing, overcomes the inertia of the actuator, and moves the actuator to do work. Friction uses up a portion of the energy as the fluid flows through the lines and components.

RELATIONSHIP OF FORCE, PRESSURE, AND HEAD

In dealing with fluids, forces are usually considered in relation to the areas over which they are applied. As previously discussed, a force acting over a unit area is a pressure, and pressure can be stated either in pounds per square inch or in terms of head, which is the vertical height of the column of fluid whose weight would produce that pressure.

All five of the factors that control the actions of fluids can be expressed either as force or in terms of equivalent pressures or head. In either situation, the different factors are referred to in the same terms.

STATIC AND DYNAMIC FACTORS

Gravity, applied forces, and atmospheric pressure are examples of static factors that apply equally to fluids at rest or in motion. Inertia and friction are dynamic forces that apply only to fluids in motion. The mathematical sum of gravity, applied forces, and atmospheric pressure is the static pressure obtained at any one point in a fluid system at a given point in time. Static pressure exists in addition to any dynamic factors that may also be present at the same time.

Remember that Pascal's law states that a pressure set up in a fluid acts equally in all directions and at right angle to the containing surfaces. This covers the situation only for fluids at rest. It is true only for the factors making up static head. Obviously, when velocity becomes a factor, it must have a direction of flow. The same is true of the force created by velocity. Pascal's law alone does not apply to the dynamic factors of fluid power systems.

The dynamic factors of inertia and friction are related to the static factors. Velocity head and friction heads are obtained at the expense of static head. However, a portion of the velocity head can always be reconverted to static head. Force, which can be produced by pressure or head when dealing with fluids, is necessary to start a body moving if it is at rest, and is present in some form when the motion of the body is arrested. Therefore, whenever a fluid is given velocity, some part of its original static head is used to impart this velocity, which then exists as velocity head.

BERNOULLI'S PRINCIPLE

Review the system illustrated in Figure 2-6. The chamber on the left is under pressure and is connected by a tube to the chamber on the right, which is also under pressure. The pressure in chamber A is static pressure of 100 psi. The pressure at some point along the connecting tube consists of a velocity pressure of 10 psi exerted in a direction parallel to the line of flow, plus the unused static pressure of 90 psi. The static pressure (90 psi) follows Pascal's law and exerts equal pressure in all directions. As the fluid enters the chamber on the right, it slows down and its velocity is reduced. As a volume of liquid moves from a small, confined space into a larger area, the fluid will expand to fill the greater volume. The result of this expansion is a reduction of velocity and a momentary reduction in pressure.

In the example, the force required to absorb the fluid's inertia equals the force required to start the fluid moving originally, so that the static pressure in the right-

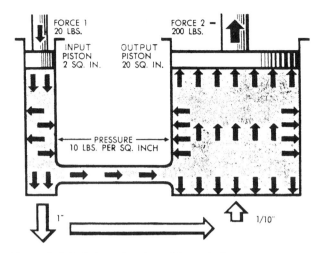

Figure 2–6 Relation of static and dynamic factors.

hand chamber is equal to that in the left-hand chamber. This example disregards friction. Therefore, it would not be encountered in actual practice. Force or head is also required to overcome friction. Unlike inertia, this force cannot be recovered. Even though the energy required to overcome friction still exists, it has been converted to heat. In an actual system, the pressure in the right-hand chamber would be less than that in the left-hand chamber. The difference would be the amount of pressure used to overcome friction within the system.

At all points in a system, the static pressure is always equal to the original static pressure less any velocity head at a specific point in the system and less the friction head required to reach that point. Since both the velocity head and friction head represent energy and energy cannot be destroyed, the sum of the static head, the velocity head, and the friction head at any point in the system must add up to the original static head. This is known as Bernoulli's principle, which states: For the horizontal flow of fluids through a tube, the sum of the pressure and the kinetic energy per unit volume of the fluid is constant. This principle governs the relationship of the static and dynamic factors in hydraulic systems.

MINIMIZING FRICTION

Fluid power equipment is designed to reduce friction as much as possible. Since energy cannot be destroyed, some of the energy created by both static pressure and velocity is converted to heat energy as the fluid flows through the piping and components within a hydraulic system. As friction increases, so does the amount of dynamic and static energy that is converted into heat.

To minimize the loss of useable energy lost to its conversion to heat energy, care must be taken in the design, installation, and operation of hydraulic systems. As a minimum, the following factors must be considered:

Proper fluid must be chosen and used in the system. It must have the viscosity, operating temperature range, and other characteristics that are conducive to proper operation of the system and to the lowest possible friction component.

Fluid flow is also critical for proper operation of a hydraulic system. Turbulent flow should be avoided as much as possible. Clean, smooth pipe or tubing should be used to provide laminar flow and the lowest friction possible within the system. Sharp, close-radius bends and sudden changes in cross-sectional areas are avoided.

System components, such as pumps, valves, and gauges, create both turbulent flow and high friction. Pressure drop, or a loss of pressure, is created by a combination of turbulent flow and friction as the fluid flows through the unit. System components that are designed to provide minimum interruption of flow and pressure should be selected for the system.

TRANSMISSION OF FORCE THROUGH LIQUIDS

When the end of a solid bar is struck, the main force of the blow is carried straight through the bar to the other end (Figure 2-7, view A). This happens because the bar is rigid. The direction of the blow almost entirely determines the direction of the transmitted force. The more rigid the bar, the less force is lost inside the bar or transmitted outward at right angles to the direction of the blow.

When a force is applied to the end of a column of confined liquid (Figure 2-7, view B), it is transmitted straight through to the other end. It is also equal and undiminished in every direction throughout the column—forward, backward, and sideways—so that the containing vessel is literally filled with the added pressure.

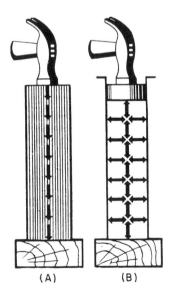

(A)   (B)

Figure 2–7 Transmission of force: (A) solid; (B) fluid.

So far we have explained the effects of atmospheric pressure on liquids and how external forces are distributed through liquids. Let us now focus our attention on forces generated by the weight of liquids themselves. To do this, we must first discuss density, specific gravity, and Pascal's law.

PRESSURE AND FORCE IN HYDRAULIC SYSTEMS

According to Pascal's law, any force applied to a confined fluid is transmitted uniformly in all directions throughout the fluid regardless of the shape of the container. Consider the effect of this in the system shown in Figure 2-8.

If there is a resistance on the output piston and the input piston is pushed downward, a pressure is created through the fluid, which acts equally at right angles to surfaces in all parts of the container. If force 1 is 100 pounds and the area of the input piston is 10 square inches, then the pressure in the fluid is 10 psi.

$$\frac{100 \text{ lbs}}{10 \text{ square inches}}$$

Note: Fluid pressure cannot be created without resistance to flow. In this case, the equipment to which the output piston is attached provides resistance. The force of resistance acts against the top of the output piston. The pressure is created in the system by the input piston pushing on the underside of the output piston with a force of 10 pounds per square inch.

In this case, the fluid column has a uniform cross-section, so the area of the output piston is the same as the area of the input piston, or 10 square inches. Therefore, the upward force on the output piston is 100 pounds and is equal to the force applied to the input piston. All that was accomplished in this system was to transmit the 100 pounds of force around the bend. However, this principle underlies practically all mechanical applications of hydraulics or fluid power.

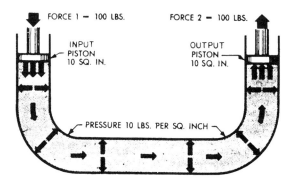

Figure 2–8 Force transmitted through fluid.

At this point, you should note that since Pascal's law is independent of the shape of the container, it is not necessary that the tube connecting the two pistons have the same cross-sectional area as the pistons. A connection of any size, shape, or length will do, as long as an unobstructed passage is provided. Therefore, the system shown in Figure 2-9, with a relatively small, bent pipe connecting two cylinders, will act exactly the same as the system shown in Figure 2-8.

MULTIPLICATION OF FORCE

Unlike the preceding discussion, hydraulic systems can provide mechanical advantage or a multiplication of input force. Figure 2-10 illustrates an example of an increase in output force. Assume that the area of the input piston is 2 square inches. With a resistant force on the output piston, a downward force of 20 pounds acting on the input piston will create a pressure of 20/2 or 10 psi in the fluid. Although this force is much smaller than the force applied in Figures 2-8 and 2-9, the pressure is the same. This is because the force is applied to a smaller area.

This pressure of 10 psi acts on all parts of the fluid container, including the bottom of the output piston. The upward force on the output piston is 200 pounds (10 psi × piston area). In this case, the original force has been multiplied tenfold while using the same pressure in the fluid as before. In any system with these dimensions, the ratio of output force to input force is always 10 to 1, regardless of the applied force. For example, if the applied force of the input piston is 50 pounds, the pressure in the system will be 25 psi. This will support a resistant force of 500 pounds on the output piston. The system works the same in reverse.

If we change the applied force and place a 200-pound force on the output piston (Figure 2-11), making it the input piston, the output force on the input piston will be one-tenth the input force, or 20 pounds. Therefore, if two pistons are used in a fluid power system, the force acting on each piston is directly proportional to its area, and the magnitude of each force is the product of the pressure and the area of each piston.

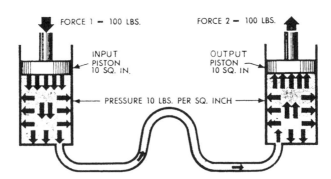

Figure 2–9 Transmitting force through small pipe.

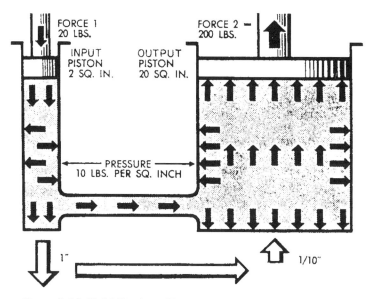

Figure 2–10 Multiplication of forces.

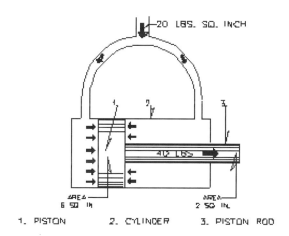

Figure 2–11 Differential areas on a piston.

DIFFERENTIAL AREAS

Figure 2-11 is a simple example of differential pressure. The figure illustrates a single piston with a surface area of 6 square inches, attached to a piston rod with an area of 2 square inches. Without any external force applied to the end of the piston rod, an equal force, 20 psig, applied to both sides of the piston, will cause the piston to move to the right. This motion is the result of differential forces. Even though the input force, 20 psig, is applied to both sides of the piston, the difference in area, 6 in^2 on the left face, and 2 in^2 on the right face, will cause the piston to move. The opposed faces of the piston behave like two pistons acting against each other. The area of one face is the full cross-sectional area of the cylinder or 6 square inches, while the area of the opposing face is the area of the piston minus the area of the piston rod, or 2 square inches. This leaves an effective area of 4 square inches on the piston rod side of the piston.

The force acting on the left side of the piston is equal to 20 psi × 6 square inches or 120 pounds. The opposing force generated by the right side of the piston is 20 psi × 4 square inches or 80 pounds. Therefore, there is a net unbalanced force of 40 pounds (120 – 80) acting on the right, and the piston will move in that direction.

VOLUME AND DISTANCE FACTORS

You have learned that if a force is applied to a system and the cross-sectional areas of the input and output are equal, the force on the input piston will support an equal resistant force on the output piston. The pressure of the liquid at this point is equal to the force applied to the input piston divided by the piston's area. Let us now look at what happens when a force greater than the resistance is applied to the input piston.

In the system illustrated in Figure 2-8, assume that the resistant force on the output piston is 100 pounds. If a force slightly greater than 100 pounds is applied to the input piston, the pressure in the system will be slightly greater than 10 psi. This increase in pressure will overcome the resistant force on the output piston. If the input piston is forced downward 1 inch, the movement displaces 10 cubic inches of fluid. The fluid must go somewhere. Since the system is closed and the fluid is practically incompressible, the fluid will move the right side of the system. Because the output piston also has a cross-sectional area of 10 square inches, it will move upward 1 inch to accommodate the 10 cubic inches of fluid. You may generalize this by saying that if two pistons in a closed system have equal cross-sectional areas and one piston is pushed and moved, the other piston will move the same distance in the opposite direction. This is because a decrease in volume in one part of the system is balanced by an equal increase in volume in another part of the system.

Apply this reasoning to the system in Figure 2-9. If the input piston is pushed down a distance of 1 inch, the volume in the left cylinder will decrease by 2 cubic inches. At the same time, the volume in the right cylinder will increase by 2 cubic inches. Since the diameter of the right cylinder cannot change, the piston must move upward to allow the volume to increase. The piston will move a distance equal to the volume increase divided by the surface area of the piston. In this example, the piston will move one-tenth of an inch (2 cubic inches/20 square inches).

This leads to the second basic rule for fluid power systems that contain two pistons: The distances the pistons move are inversely proportional to their areas. Or, more simply, if one piston is smaller than the other, the smaller piston must move a greater distance than the larger piston any time the pistons move.

3

HYDRAULIC PUMPS

The purpose of a hydraulic pump is to supply the flow of fluid required by a hydraulic system. The pump does not create system pressure. System pressure is created by a combination of the flow generated by the pump and the resistance to flow created by friction and restrictions within the system.

As the pump provides flow, it transmits a force to the fluid. When the flow encounters resistance, this force is changed into pressure. Resistance to flow is the result of a restriction or obstruction in the flow path. This restriction is normally the work accomplished by the hydraulic system, but there can also be restrictions created by the lines, fittings, or components within the system. Thus, the load imposed on the system or the action of a pressure-regulating valve controls the system pressure.

OPERATION

A pump must have a continuous supply of fluid available to its inlet port before it can supply fluid to the system. As the pump forces fluid through the outlet port, a partial vacuum or low-pressure area is created at the inlet port. When the pressure at the inlet port of the pump is lower than the atmospheric pressure, the atmospheric pressure acting on the fluid in the reservoir must force the fluid into the pump's inlet. This is called a suction lift condition.

PERFORMANCE

Pumps are normally rated by their volumetric output and discharge pressure. Volumetric output is the amount of fluid a pump can deliver to its outlet port in a certain period of time and at a given speed. Volumetric output is usually expressed in gallons per minute (gpm).

Since changes in pump speed affect volumetric output, some pumps are rated by their displacement. Pump displacement is the amount of fluid the pump can deliver per cycle or complete rotation. Since most pumps use a rotary drive, displacement is usually expressed in terms of cubic inches per revolution.

Although pumps do not directly create pressure, the system pressure created by the restrictions or work performed by the system has a direct effect on the volumetric output of the pump. As the system pressure increases, the volumetric output of the pump decreases. This drop in volumetric output is the result of an increase in the amount of leakage within the pump. This leakage is referred to as pump slippage or slip. It is a factor that must be considered in all hydraulic pumps.

PUMP RATINGS

Pumps are generally rated by their maximum operating pressure capability and their output in gallons per minute (gpm) at a given operating speed.

Pressure

Based on reasonable service life expectancy under specified operating conditions, the manufacturer determines the pressure rating of a pump. It is important to note that there is no standard industry-wide safety factor in this rating. Operating at higher pressure may result in reduced pump life or more serious damage.

Displacement

The flow capacity of a pump can be expressed as its displacement per revolution or by its output in gallons per minute (gpm). Displacement is the volume of liquid transferred in one complete cycle of pump operation. It is equal to the volume of one pumping chamber multiplied by the number of chambers that pass the outlet during one complete revolution or cycle. Displacement is expressed in cubic inches per revolution.

Most pumps that are used in hydraulic applications have a fixed displacement, which cannot be changed except by replacing certain components. However, in some it is possible to vary the size of the pumping chamber and thereby the displacement by means of external controls. Some unbalanced vane pumps and many piston units can be varied from maximum to zero delivery or even to reverse flow without modification to the pump's internal configuration.

Volumetric Efficiency

In theory, a pump delivers an amount of fluid equal to its displacement each cycle or revolution. In reality, the actual output is reduced because of internal leakage or slippage. As pressure increases, the leakage from the outlet to the inlet or to the drain also increases and the volumetric efficiency decreases.

Volumetric efficiency is equal to the actual output divided by the theoretical output. It is expressed as a percentage:

$$\text{Efficiency} \;=\; \frac{\text{Actual output}}{\text{Theoretical output}} \times 100$$

For example, if a pump theoretically should deliver 10 gpm but delivers only 9 gpm at 1,000 psig, its volumetric efficiency at that pressure is 90 percent.

$$\text{Efficiency} \;=\; \frac{9 \text{ gpm}}{10 \text{ gpm}} \times 100 \;=\; 90 \text{ \%}$$

If the discharge pressure is increased, the amount of slippage will increase. If we increase the pressure in the preceding example to 1,500 psig, the actual output may drop to 8 gpm. Therefore, the volumetric efficiency will decrease to 80 percent at 1,500 psig.

PUMP CLASSIFICATIONS

Many different methods are used to classify pumps. Terms such as hydrodynamic, positive-displacement, fixed-displacement, variable-displacement, constant-volume, and others are used to describe hydraulic pumps.

Positive-displacement pumps, unlike centrifugal pumps, will provide a definite volume of fluid for each cycle of pump operation, regardless of the resistance offered by the system, provided the capacity of the power unit driving the pump is not exceeded. If the outlet of a positive-displacement pump were completely closed, the pressure would instantaneously increase to the point at which the pump driver would stall or something in the drive-train would break.

Positive-displacement pump classification can be subdivided into other classifications that include fixed-displacement or variable-displacement. Other terms, such as fixed-delivery, constant-delivery, and constant-volume, may be used to describe this type of pump.

The fixed-displacement pump delivers the same amount of fluid on each cycle. Only changing the speed of the pump can change the output volume. When a pump of this type is used in a hydraulic system, a pressure regulator or relief valve must be installed in the system.

The variable-displacement classification of hydraulic pumps is constructed so that the displacement per cycle can be varied. The displacement is varied through the use of an internal control device. The construction of these devices can vary from an unloading or pressure regulating valve to restricted-flow bypass loops. Some of these devices will be described in the control valve section.

Hydraulic pumps may also be classified according to the specific design used to create the flow of fluids. Practically all hydraulic pumps fall within three design classifications: centrifugal, rotary, and reciprocating. The use of centrifugal pumps in hydraulics is extremely limited and will not be discussed in this text.

Hydrodynamic Pumps

Hydrodynamic or non-positive-displacement pumps such as centrifugal or turbine designs are used primarily in the transfer of fluids where the only resistance encountered is that created by the weight of the fluid itself and friction.

Most non-positive-displacement pumps (Figure 3-1) operate by centrifugal force where fluids entering the center of the pump housing are thrown to the outside by a rapidly driven impeller. There is no positive seal between the inlet and outlet ports, and pressure capabilities are a function of rotating speed.

Although they provide smooth continuous flow, the output of a hydrodynamic pump is reduced as resistance is increased. It is possible to completely block off or *deadhead* the outlet while the pump is running. For this and other reasons, non-positive-displacement pumps are seldom used in hydraulic systems.

Hydrostatic Pumps

Hydrostatic or positive-displacement pumps, as their name implies, provide a given amount of fluid for every stroke, revolution, or cycle. Except for leakage, their output is independent of outlet pressure or back pressure from the system. This properly makes them well suited for use in the transmission of power.

Rotary Pumps

All rotary pumps have rotating parts, which trap the fluid at the inlet port and force it, through the discharge port, into the system. Gears, screws, lobes, and vanes are com-

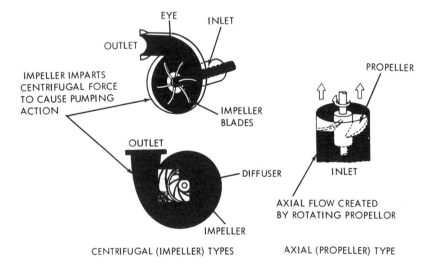

Figure 3–1 Non-positive-displacement pumps.

monly used to move the fluid within the pump. Rotary pumps are classified as positive, fixed-displacement type.

Rotary pumps are designed with very small clearances between their rotating and stationary parts to minimize slippage from the discharge side back to the suction side of the pump. They are designed to operate at relatively moderate speeds, normally below 1,800 rpm. Operation at higher speeds can cause erosion and excessive wear.

There are numerous types of rotary pumps and various methods of classification. They may be classified by the shaft position, the type of driver, their manufacturer's name, or their service application. However, classification of rotary pumps is generally made according to the type of rotating element. A few of the more common types include the following.

Gear Pumps

A gear pump develops flow by carrying fluid between the teeth of two meshed gears. One gear is driven by the drive shaft and turns the other. The pumping chambers formed between the gear teeth are enclosed by the pump's housing and the side plates.

A partial vacuum is created at the inlet as the gear teeth unmesh. Fluid flows in to fill the space and is carried around the outside of the gears. As the teeth mesh again at the outlet, the fluid is forced out. High pressure at the pump's outlet imposes an unbalanced load on the gears and their bearing support structure.

Gear pumps are classified as either external or internal gear pumps. In external gear pumps, the teeth of both gears project outward from their centers (Figure 3-2). External gear pumps may use spur, herringbone, or helical gear sets to move the fluid.

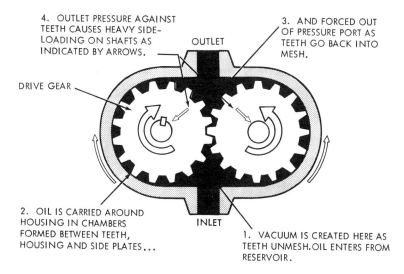

Figure 3–2 External gear rotary pump.

External Gear Pumps

In this design, the pumping chambers also are formed between the gear teeth. A crescent seal is machined into the pump body between the inlet and outlet where clearance between the teeth is at its greatest.

Also in this general family of gear pumps is the lobe or rotor pump. This pump operates on the same principle as the external gear, but has a higher displacement.

Spur Gear Pumps

The spur gear pump, shown in Figure 3-2, consists of two meshed gears, which revolve in a housing. The drive gear in the illustration is turned by a drive shaft, which is attached to the power source. The clearances between the gear teeth as they mesh and between the teeth and the pump housing are very small.

In Figure 3-2, the drive gear is turning in a counterclockwise direction and the driven gear is turning in a clockwise direction. As the teeth pass the inlet port, liquid is trapped between the teeth and the housing. This liquid is carried around the housing to the outlet port. As the teeth mesh again, the liquid between the teeth is pushed into the outlet port. This action produces a positive flow of liquid into the system. A shear pin or shear section is incorporated in the drive shaft. This is to protect the power source or reduction gears if the pump fails because of excessive load or the jamming of parts.

Herringbone Gear Pumps

The herringbone gear pump (Figure 3-3) is a modification of the spur gear design. The liquid is pumped in the same manner as in the spur gear pump. However, in the

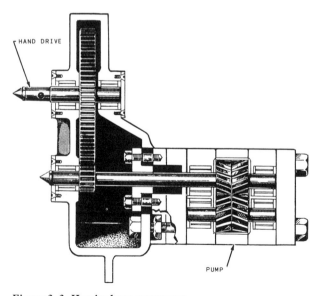

Figure 3–3 Herringbone gear pump.

herringbone pump, each set of teeth begins its fluid discharge phase before the preceding set of teeth has completed its discharge phase. This overlapping and the relatively larger space at the center of the gears tend to minimize pulsations and give a steadier flow than the spur gear pump.

Helical Gear Pumps

The helical gear pump (Figure 3-4) is still another modification of the spur gear design. Because of the helical gear design, the overlapping of successive discharges from spaces between the teeth is even greater than it is in the herringbone design. Therefore, the discharge flow is smoother. As a result, the gears can be designed with a small number of large teeth, thus allowing an increase in capacity without sacrificing smooth flow.

The gear sets in this type of pump are driven by a set of timing gears that help maintain the required close tolerance between mating gears without actual metal-to-metal contact. Metallic contact between the teeth would provide a tighter seal against hydraulic slip, but it would also decrease volume and dramatically increase wear of the teeth.

Anti-friction bearings at both ends of the gear shafts maintain proper alignment and radial clearance between gears and minimize the friction loss in the transmitted power. Suitable packing is used to prevent leakage around the shaft.

Internal Gear Pumps

In an internal gear pump, the teeth of one gear project outward from the gear hub, the teeth of the other gear projects inward toward the center of the pump (Figure 3-5, view A). Internal gear pumps may be either centered or off-centered. Figure 3-5 illustrates two types of internal gear pumps. Pump A is an example of a centered pump, and pump B is off-centered.

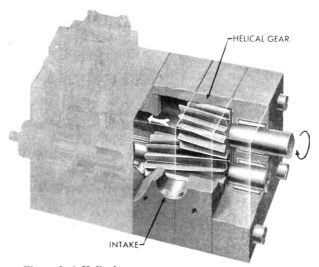

Figure 3–4 Helical gear pump.

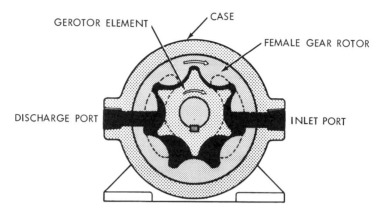

Figure 3–5 Internal gear pumps.

Off-Centered Internal Gear Pumps

In this type of pump, the drive gear is attached directly to the drive shaft of the pump and is placed off-center in relation to the internal gear. The two gears mesh on one side of the pump, between the suction and discharge ports. On the opposite side of the chamber, a crescent-shaped form fitted to a close tolerance fills the space between the two gears.

The rotation of the center gear by the drive shaft causes the outside gear to rotate, since the two gears are meshed. Everything in the chamber rotates except the crescent. This causes liquid to be trapped in the gear spaces as they pass the crescent. The liquid is carried from the suction port to the discharge port, where it is forced out of the pump by the meshing gears. The size of the crescent that separates the internal and external gears determines the volume delivered by the pump. A small crescent allows more volume of liquid per revolution than a larger crescent.

Centered Internal Gear

Another design of internal gear pump is illustrated in Figures 3-6 and 3-7. This pump consists of a pair of gear-shaped elements, one within the other, located in the pump chamber. The inner gear is connected to the drive shaft of the power source.

The operation of this type of internal gear pump is illustrated in Figure 3-7. To simplify the explanation, the teeth of the inner gear and the space between teeth of the outer gear are numbered. Note that the inner gear has one tooth fewer than the outer gear. The tooth forms of the two gears are related in such a way that each tooth of the inner gear is always in sliding contact with the surface of the outer gear. Each tooth of the inner gear meshes with the outer gear at just one point during each revolution. In the illustration, this point is at the (X). In view A, tooth 1 of the inner gear is meshed with space 1 of the outer gear. As the gears continue to rotate in a clockwise direction

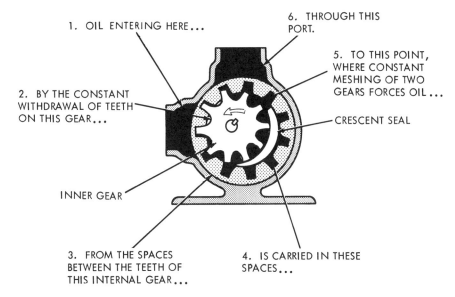

1. OIL ENTERING HERE...

6. THROUGH THIS PORT.

5. TO THIS POINT, WHERE CONSTANT MESHING OF TWO GEARS FORCES OIL...

2. BY THE CONSTANT WITHDRAWAL OF TEETH ON THIS GEAR...

CRESCENT SEAL

INNER GEAR

3. FROM THE SPACES BETWEEN THE TEETH OF THIS INTERNAL GEAR...

4. IS CARRIED IN THESE SPACES...

Figure 3–6 Centered internal gear pump.

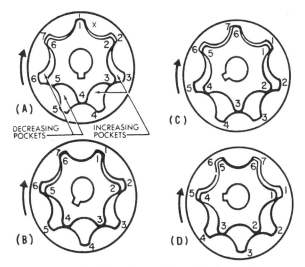

(A)

DECREASING POCKETS

INCREASING POCKETS

(B)

(C)

(D)

Figure 3–7 Principles of operation of the internal gear pump.

and the teeth approach point X, tooth 6 of the inner gear will mesh with space 7 of the outer gear, tooth 5 with space 6, and so on. During this revolution, tooth 1 will mesh with space 2; and during the following revolution with space 3. As a result, the outer gear will rotate at just six-sevenths the speed of the inner gear.

At one side of the point of mesh, pockets of increasing size are formed as the gears rotate, while on the other side the pockets decrease in size. In Figure 3-7, the pockets on the right-hand side of the drawings are increasing in size toward the bottom of the illustration, while those on the left-hand side are decreasing in size toward the top of the illustration. The intake side of the pump would therefore be on the right and the discharge side on the left. In Figure 3-7, since the right-hand side of the drawing was turned over to show the ports, the intake and discharge appear reversed. Actually, A in one drawing covers A in the other.

Lobe Pumps

The lobe pump uses the same principles of operation as the external gear pump. The lobes are considerably larger than gear teeth, but there are only two or three lobes on each rotor. A three-lobed pump is illustrated in Figure 3-8. The two elements are rotated, one directly driven by the power source, and the other through timing gears. As the elements rotate, liquid is trapped between two lobes of each rotor and the walls of the pump chamber. The trapped liquid is carried from the suction side to the discharge side of the pump chamber. As liquid leaves the suction chamber, the pressure in the suction chamber is lowered and additional liquid is pulled into the chamber from the reservoir.

The lobes are constructed so there is a continuous seal at the points where the two lobes meet at the center of the pump. The lobes of the pump illustrated in Figure 3-8 are fitted with small vanes at the outer edge to improve the seal of the pump. Although these vanes are mechanically held in their slots, they are free to move outward. Centrifugal force keeps the vanes snug against the chamber and the other rotating members.

Vane Pumps

Vane-type hydraulic pumps generally have circularly or elliptically shaped interiors and flat end plates. Figure 3-9 illustrates a vane pump with a circular interior. A slotted rotor is fixed to a shaft that enters the pump-housing cavity through one of its end plates. A number of small rectangular plates or vanes are set into the slots of the rotor. As the rotor turns, centrifugal force causes the outer edge of each vane to slide along the surface of the

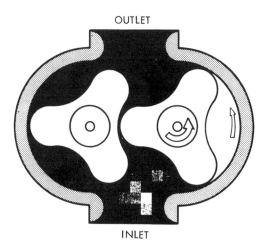

Figure 3–8 Lobe pump.

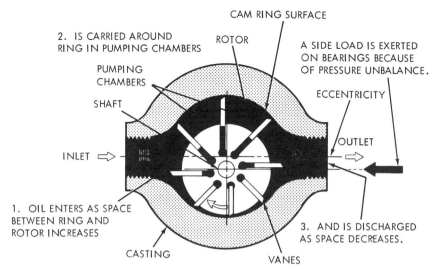

Figure 3–9 Vane pump.

housing cavity. The cavities formed by the vanes, the end plates, the housing, and the rotor enlarge and shrink as the rotor and vane assembly rotates. An inlet port is installed in the housing so fluid may flow into the cavities as they enlarge. An outlet port is provided to allow the fluid to flow out of the cavities, as they become small.

The pump shown in Figure 3-9 is referred to as an unbalanced pump because all of the pumping action takes place on one side of the rotor. This causes a side load on the rotor. Some vane pumps are constructed with an elliptical housing that forms two separate pumping areas on opposite sides of the rotor. This cancels the side load. This type of pump is referred to as a balanced vane.

Usually, vane pumps are fixed-displacement and pump only in one direction. There are some designs of vane pumps that provide variable flow. Vane pumps are generally restricted to service where pressure demand does not exceed 2,000 psi. Wear rates, vibration, and noise level increase rapidly in vane pumps as pressure demands exceed 2,000 psi.

RECIPROCATING PUMPS

The term reciprocating is defined as back-and-forth motion. In the reciprocating pump it is the back-and-forth motion of pistons inside of cylinders that provides the flow of fluid. Reciprocating pumps, like rotary pumps, operate on the positive principle: that is, each stroke delivers a definite volume of liquid to the system.

One major limitation of reciprocating pumps is the intermittent flow that they produce. The back-and-forth motion generates pulses of volume that create vibration and turbulent flow within the hydraulic system. These systems must include an accumulator downstream from the pump to dampen these pulses.

Hand Pumps

There are two types of manually operated reciprocating pumps—the single-action and the double-action. The single-action pump provides flow during every other stroke, while the double-action provides flow during each stroke. Single-action pumps are frequently used in hydraulic jacks.

A double-action hand pump is illustrated in Figure 3-10. This type of pump is used in some applications as a source of emergency hydraulic power or for testing hydraulic systems.

This type of pump consists of a cylinder, a piston containing a built-in check valve (A), a piston rod, an operating handle, and a check valve (B) at the inlet port. When the piston is moved to the left, the force of the liquid in the outlet chamber and spring tension cause valve A to close. This movement causes the piston to force the liquid in the outlet chamber through the outlet port and into the system. This same piston movement causes a low-pressure area in the inlet chamber and the liquid, at atmospheric pressure, in the reservoir acting on check valve B causes its spring to compress. Thus, opening the check valve allows liquid to enter the inlet chamber.

When the piston completes this stroke to the left, the inlet chamber is full of liquid. This eliminates the pressure difference between the inlet chamber and the reservoir, thereby allowing spring tension to close check valve B. When the piston is moved to the right, the force of the confined liquid in the inlet chamber acts on check valve A. This action compresses the spring and opens the valve, allowing the liquid to flow from the intake chamber to the outlet chamber. Because of the area occupied by the piston rod, the outlet chamber cannot contain all the liquid discharged from the inlet chamber. Since liquids do not compress, the extra liquid is forced out of the outlet port and into the system.

Piston Pumps

All piston pumps operate on the principle that a piston reciprocating in a bore will draw in fluid as it is retracted and expel it on the forward stroke. Two basic designs are radial and axial, both are available as fixed- or variable-displacement models. A radial pump has the pistons arranged radially or at 90 degrees to the centerline of the drive shaft (Figure 3-11). In an axial configuration (Figure 3-12), the pistons are parallel to each other and to the axis of the cylinder block. The latter may be further divided into in-line and bent axis types.

A further distinction is made between pumps that provide a fixed delivery and those able to vary the flow of the hydraulic fluid. Variable-delivery pumps can be further

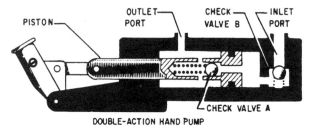

Figure 3-10 Hydraulic hand pump.

divided into those able to pump fluid from zero to full delivery in one direction of flow and those able to pump from zero to full delivery in either direction.

Radial Piston Pumps

In a radial pump, the cylinder block rotates on a stationary pintle and inside a circular reaction ring or rotor. As the block rotates, centrifugal force, charging pressure, or some form of mechanical action causes the pistons to follow the inner surface of the ring, which is offset from the centerline of the cylinder block. As the pistons recipro-cate in their bores, porting in the pintle permits them to take in fluid as they move out-ward and discharge it at a higher pressure as they move in.

The size and number of pistons and the length of their stroke determines pump dis-placement. In some models, moving the reaction ring to increase or decrease the pis-ton travel length or stroke can vary the displacement.

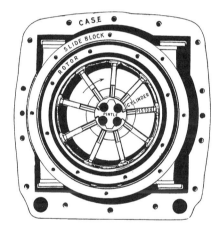

Figure 3–11 Radial piston pump.

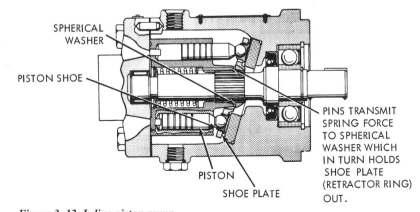

Figure 3–12 Inline piston pump.

Figure 3-13 illustrates the operation of the radial piston pump. The pump consists of a stationary pintle that acts as a valve and a cylinder block, which revolves around the pintle. The cylinder block also contains the pistons; a rotor that houses the reaction ring against which the piston heads press; and a slide block that is used to control the length of piston stroke. The slide block does not revolve, but houses and supports the rotor, which does revolve because of the friction set up by the sliding action between the piston heads and the reaction ring. The cylinder block is attached to the drive shaft.

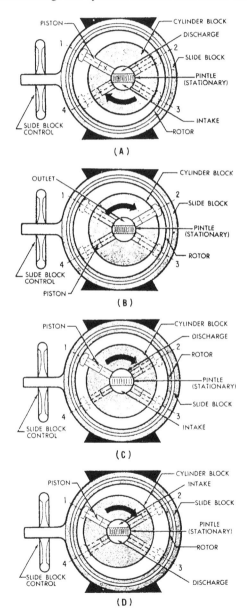

Figure 3–13 Operation of the radial piston pump.

Referring to view A of Figure 3-13, assume that space X in one of the cylinders of the cylinder block contains fluid and that the respective piston of this cylinder is at position 1. When the cylinder block and piston are rotated in a clockwise direction, the piston is forced into its cylinder as it approaches position 2. This action reduces the volumetric size of the cylinder and forces a quantity of fluid out of the cylinder and into the outlet port above the pintle. This pumping action is due to the rotor being off-center in relation to the center of the cylinder block.

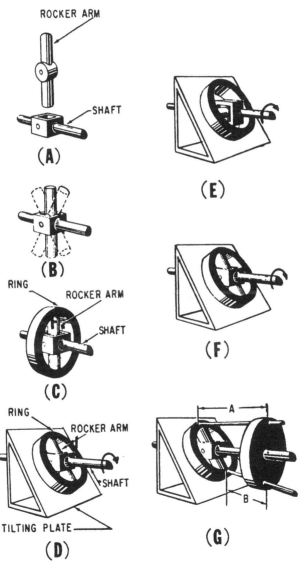

Figure 3–14 Relationship of the universal joint in operation of an axial piston pump.

In Figure 3-13, view B, the piston has reached position 2 and has forced the fluid out of the open end of the cylinder, through the outlet above the pintle and into the system. While the piston moves from position 2 to position 3, the open end of the cylinder passes over the solid part of the pintle; therefore, there is no intake or discharge of fluid. As the piston and cylinder move from position 3 to position 4, centrifugal force causes the piston to move outward against the reaction ring of the rotor. During this time the open end of the cylinder is open to the intake side of the pintle and fills with fluid. As the piston moves from position 4 to position 1, the open end of the cylinder is against the solid side of the pintle and no intake or discharge of fluid takes place. After the piston has passed the pintle and starts toward position 2, another discharge of fluid takes place. Alternate intake and discharge continues as the rotor revolves about its axis—intake on one side of the pintle and discharge on the other.

Notice in views A and B of Figure 3-13 that the center point of the rotor is different from the center point of the cylinder block. The difference of these centers produces the pumping action. If the rotor is moved so that its center point is the same as that of the cylinder block, as shown in Figure 3-13, view C, there is no pumping action. Since the piston does not move back and forth in the cylinders as it rotates within the cylinder block, no pumping can take place.

The flow in this pump can be reversed by moving the slide block and rotor to the right so the relation of the centers of the rotor and cylinder block is reversed from the position shown in views A and B of Figure 3-13. View D shows this arrangement. Fluid enters the cylinder as the piston travels from position 1 to position 2 and is discharged from the cylinder as the piston travels from position 3 to position 4.

In the illustrations, the rotor is shown in the center, the extreme right, or the extreme left as related to the cylinder block. The amount of adjustment in distance between the two centers determines the length of the piston stroke and the amount of fluid flow in and out of the cylinder. Thus, the adjustment determines the displacement of the pump. This adjustment may be controlled in different ways. Manual control by a handwheel is the simplest. The pump illustrated in Figure 3-13 is controlled in this way. For automatic control of delivery to accommodate varying volume requirements during the operating cycle, a hydraulically controlled cylinder may be used to position the slide block. A gear motor controlled by a push button or a limit switch is sometimes used for this purpose.

Swash Plate Design Pumps

In axial piston pumps, the cylinder block and drive shaft are on the same centerline and the pistons reciprocate parallel to the drive shaft. The simplest type of axial piston pump is the swash plate inline design (Figure 3-15).

The cylinder blow in this pump is turned by the drive shaft. Pistons fitted to bores in the cylinder are connected through piston shoes and a retracting ring, so that the shoes bear against an angled swash plate. As the block turns (Figure 3-16), the piston shoes follow the swash plate, causing the pistons to reciprocate. The ports are arranged in

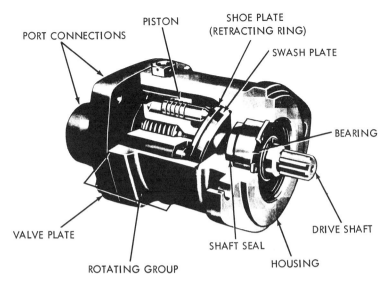

Figure 3–15 Inline design piston pump.

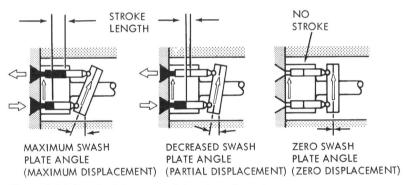

Figure 3–16 Swash plate causes pistons to reciprocate.

the valve plate so that the pistons pass the inlet as they are being pulled out and pass the outlet as they are being forced forward.

In these pumps the size and number of pistons as well as their stroke length also determine the displacement. The stroke length is controlled by the swash plate angle. In variable-displacement models, the swash plate is installed in a movable yoke (Figure 3-17). By pivoting the yoke on pintles, the swash plate angle and piston stroke can be increased or decreased. Figure 3-17 shows a compensator control, but the angle can also be controlled manually or by a variety of other means.

Operation of the inline compensator-controlled pump is shown schematically in Figure 3-17. The control consists of a compensator valve balanced between load pressure and the force of a spring, a piston controlled by the valve to move the yoke, and a yoke return

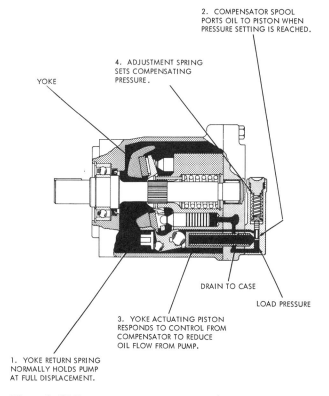

Figure 3–17 Pressure compensator control.

spring. With no outlet pressure, the yoke return spring moves the yoke to the full-delivery position. As pressure builds, it acts against the end of the valve spool. When the pressure is high enough to overcome the valve spring, the spool is displaced and oil enters the yoke piston. The piston is forced by the oil under pressure to decrease the pump displacement. If the pressure falls off, the spool automatically moves back, oil is discharged from the piston to the inside of the pump casing, and the spring returns the yoke to a greater angle.

The compensator adjusts the pump outlet to whatever displacement is required to develop and maintain the preset pressure. This prevents excess power loss by avoiding relief valve operation at full pump volume in holding and clamping applications.

Wobble Plate Inline Pumps

A variation of the inline piston design is the wobble plate pump. In a wobble plate design, the cylinder is stationary and the canted plate is turned by the drive shaft. As the plate turns, it *wobbles* and pushes against spring-loaded pistons to force them to reciprocate. Separate inlet and outlet check valves are required because the cylinders do not move past the ports.

Bent Axis Pumps

In a bent axis piston pump (Figure 3-18), the cylinder block turns with the drive shaft, but at an offset angle. The piston rods are attached to the drive shaft flange by ball joints and are forced in and out of their bores as the distance between the drive shaft flange and cylinder block changes (Figure 3-19). A universal link keys the cylinder block to the drive shaft to maintain alignment and ensure that they turn together. Except to accelerate and decelerate the cylinder block and to overcome resistance of the oil-filled housing, the link does not transmit force.

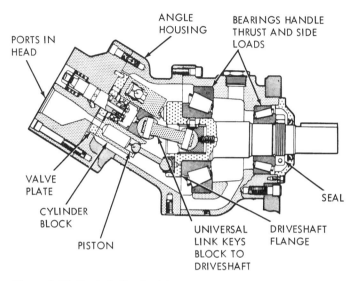

ANGLE HOUSING

BEARINGS HANDLE THRUST AND SIDE LOADS

PORTS IN HEAD

VALVE PLATE

CYLINDER BLOCK

PISTON

UNIVERSAL LINK KEYS BLOCK TO DRIVESHAFT

DRIVESHAFT FLANGE

SEAL

Figure 3–18 Bent shaft axis piston pump.

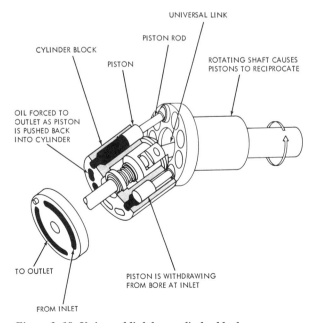

UNIVERSAL LINK

PISTON ROD

CYLINDER BLOCK

PISTON

ROTATING SHAFT CAUSES PISTONS TO RECIPROCATE

OIL FORCED TO OUTLET AS PISTON IS PUSHED BACK INTO CYLINDER

TO OUTLET

PISTON IS WITHDRAWING FROM BORE AT INLET

FROM INLET

Figure 3–19 Universal link keys cylinder block.

The displacement of this type of pump varies with the offset angle (Figure 3-20), the maximum angle being 30 degrees and the minimum zero. Fixed-displacement models (Figure 3-21) are usually available with 23-degree or 30-degree angles. In the variable-displacement construction (Figure 3-22), a yoke with an external control is used to change the angle. With some controls, the yoke can be moved over center to reverse the direction of flow.

Various methods are used to control the displacement of bent-axis pumps. Typical controls are the handwheel, pressure compensator and servo. Figure 3-23 shows a pressure compensator control for a bent-axis pump. In view A, the system pressure is

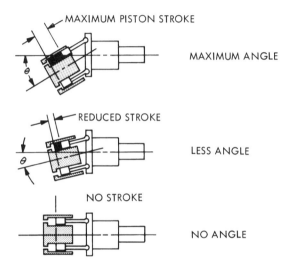

Figure 3–20 30 degrees maximum angle.

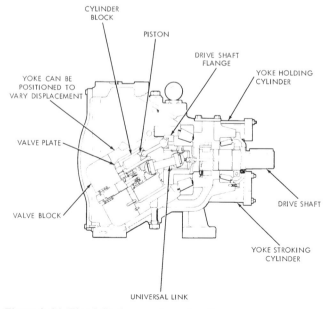

Figure 3–21 Fixed-displacement construction.

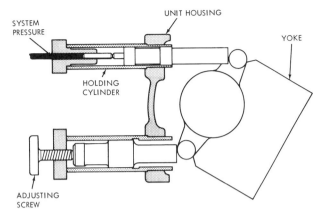

Figure 3–22 Variable-displacement construction.

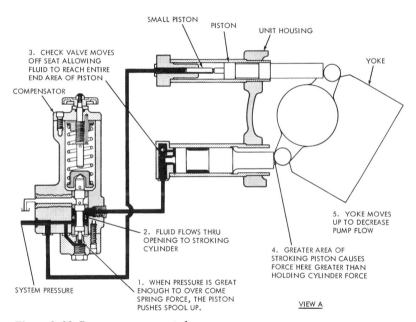

Figure 3–23 Pressure-compensated pump.

sufficient to overcome the spring force of the compensator. As a result, the spool lifts, allowing fluid to flow into the stroking cylinder. Although the holding cylinder also has system pressure applied, the area of the stroking cylinder piston is much greater. Because of the differential pressure, the yoke is forced up to decrease flow. View B shows the yoke moving down as system pressure drops below that required to overcome the compensator spring force.

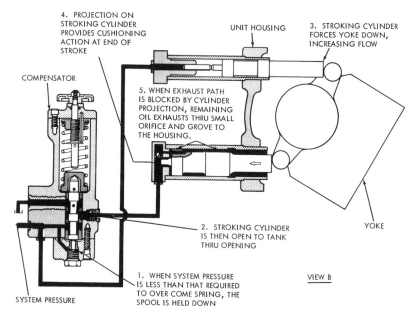

4. PROJECTION ON STROKING CYLINDER PROVIDES CUSHIONING ACTION AT END OF STROKE

UNIT HOUSING

3. STROKING CYLINDER FORCES YOKE DOWN, INCREASING FLOW

COMPENSATOR

5. WHEN EXHAUST PATH IS BLOCKED BY CYLINDER PROJECTION, REMAINING OIL EXHAUSTS THRU SMALL ORIFICE AND GROVE TO THE HOUSING.

YOKE

2. STROKING CYLINDER IS THEN OPEN TO TANK THRU OPENING

VIEW B

1. WHEN SYSTEM PRESSURE IS LESS THAN THAT REQUIRED TO OVER COME SPRING, THE SPOOL IS HELD DOWN

SYSTEM PRESSURE

Figure 3–23 Continued.

4

HYDRAULIC FLUIDS

Selection and care of the hydraulic fluid for a machine will have an important effect on how it performs and on the life of the hydraulic components. During the design of equipment that requires fluid power, many factors are considered in selecting the type of system to be used—hydraulic, pneumatic, or a combination of the two. Some of the factors required are speed and accuracy of operation, surrounding atmospheric conditions, economic conditions, availability of replacement fluid, required pressure level, operating temperature range, contamination possibilities, cost of transmission lines, limitations of the equipment, lubricity, safety to the operators, and expected service life of the equipment.

After the type of system has been selected, many of these same factors must be considered in selecting the fluid for the system. This chapter is devoted to hydraulic fluids. Included in it are sections on the properties and characteristics desired of hydraulic fluids; types of hydraulic fluids; hazards and safety precautions for working with, handling, and disposing of hydraulic liquids; types and control of contamination; and sampling.

PURPOSE OF THE HYDRAULIC FLUID

As a power transmission medium, the fluid must flow easily through lines and component passages. Too much resistance to flow creates considerable power loss. The fluid also must be as incompressible as possible so that action is instantaneous when the pump is started or a valve shifts.

Lubrication

In most hydraulic components, the hydraulic fluid provides internal lubrication. Pump elements and other wear parts slide against each other on a film of fluid (Figure 4-1).

47

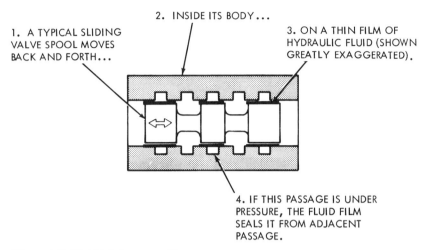

Figure 4–1 Fluid lubricates working parts.

For long component life, the oil must contain the necessary additives to ensure high anti-wear characteristics. Not all hydraulic oils contain these additives.

Sealing

In many applications, hydraulic fluid is the only seal against pressure inside system components. In Figure 4-1, there is no seal ring between the valve spool and body to prevent leakage from the high-pressure passage to the low-pressure passage. The close mechanical fit and viscosity of the hydraulic fluid determine leakage rate.

Cooling

Circulation of the hydraulic oil through lines, heat exchangers, and the walls of the reservoir (Figure 4-2) gives up heat that is generated within the system. Without this

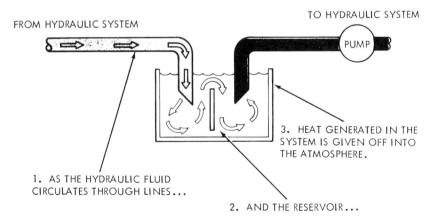

Figure 4–2 Circulating oil cools system.

cooling, the heat generated by the hydraulic pump and mechanical work performed by system actuators would build to a point that damage of system components could cause premature failure of the system.

PROPERTIES OF HYDRAULIC FLUID

If fluidity—the physical property of a substance that enables it to flow—and incompressibility were the only properties required, any liquid that is not too thick might be used in a hydraulic system. However, a satisfactory liquid for a particular system must possess a number of other properties. The most important properties and some characteristics are discussed in the following paragraphs.

Density and Specific Gravity

The density of a substance is its weight per unit of volume. The unit of volume in the English system of measurement is 1 cubic foot, or 1 ft^3. To find the density of a substance, you must know its weight and volume. You then divide its weight by its volume to find the weight per unit volume. In equation form, this is written as

$$D = \frac{W}{V}$$

EXAMPLE: The liquid that fills a certain container weighs 1,496.6 pounds. The container is 4 feet long, 3 feet wide, and 2 feet deep. Its volume is 24 cubic feet (4 ft × 3 ft × 2 ft). If 24 cubic feet of this liquid weighs 1,497.6 pounds, then 1 cubic foot weighs

$$D = \frac{1,497.6}{24}$$

or 62.4 pounds. Therefore, the density of the liquid is 62.4 pounds per cubic foot or 62.4 lb/ft^3. This is the density of water at 4°C and is usually used as the standard for comparing the densities of other substances. This standard temperature is used whenever the density of liquids and solids is measured. Changes in temperature will not change the weight of a substance but will change the volume of the substance by expansion or contraction, thus changing the weight per unit volume, density.

In physics, the word specific implies a ratio. Weight is the measure of the earth's attraction for a body, which is called gravity. Thus, the ratio of the weight of a unit volume of some substance to the weight of an equal volume of a standard substance, measured under standard pressure and temperature, is called specific gravity. The terms specific weight and specific density are also sometimes used to express this ratio:

$$\text{Specific gravity, (Sp.Gr.)} = \frac{\text{Weight of the substance}}{\text{Weight of an equal volume of water}}$$

The specific gravity of water is 1.0 or a density of 62.4 lb/ft^3. If a cubic foot of a liquid weighs 68.64 pounds, then its specific gravity is 1.1 or

$$\frac{68.64}{62.4}$$

$$\text{Sp.Gr.} = \frac{\text{Density of the substance}}{\text{Density of water}}$$

Thus, the specific gravity of the liquid is the ratio of its density to the density of water. If the specific gravity of a liquid or solid is known, the density can be obtained by multiplying its specific gravity by the density of water. For example, if a hydraulic fluid has a specific gravity of 0.8, 1 cubic foot of the fluid weighs 0.8 times the weight of water (62.4 pounds) or 49.92 pounds.

Specific gravity and density are independent of the size of the sample and depend only on the substance of which it is made.

Viscosity

Viscosity is one of the most important properties of hydraulic fluids. It is a measure of a fluid's resistance to flow. A liquid such as gasoline, which flows easily, has a low viscosity, and a liquid such as tar, which flows slowly, has a high viscosity. The viscosity of a liquid is affected by changes in temperature and pressure. As the temperature of liquid increases, its viscosity decreases. That is, a liquid flows more easily when it is hot than when it is cold. The viscosity of a liquid will increase as the pressure on the liquid increases.

A satisfactory liquid for a hydraulic system must be thick enough to give a good seal at pumps, motors, valves, and so on. These components depend on close fits for creating and maintaining pressure. Any internal leakage through these clearances results in loss of pressure, instantaneous control, and pump efficiency. Leakage losses are greater with thinner liquids (low viscosity). A liquid that is too thin will also allow rapid wearing of moving parts, or of parts that operate under heavy loads. On the other hand, if the liquid is too thick, viscosity too high, the internal friction of the liquid will cause an increase in the liquid's flow resistance through clearances of closely fitted parts, lines, and internal passages. This results in pressure drops throughout the system, sluggish operation of the equipment, and an increase in power consumption.

Measurement of Viscosity

Viscosity is normally determined by measuring the time required for a fixed volume of a fluid, at a given temperature, to flow through a calibrated orifice or capillary tube. The instruments used to measure the viscosity of a liquid are known as viscosimeters.

In decreasing order of exactness, methods of defining viscosity include absolute (poise) viscosity; kinematic viscosity in centistokes; relative viscosity in Saybolt universal seconds (SUS); and SAE numbers.

Absolute Viscosity

The resistance when moving one layer of liquid over another is the basis for the laboratory method of measuring *absolute viscosity. Poise* viscosity is defined as the force (pounds) per unit of area, in square inches, required to move one parallel surface at a speed of one centimeter per second past another parallel surface when the two surfaces are separated by a fluid film 1 centimeter thick (Figure 4-3). In the metric system, force is expressed in *dynes* and area in square centimeters. Poise is also the ratio between the shearing stress and the rate of shear of the fluid.

$$\text{Absolute Viscosity} = \frac{\text{Shear Stress}}{\text{Rate of Shear}}$$

$$1 \text{ Poise} = 1 \times \left[\frac{\text{Dyne Per Second}}{\text{Square Centimeter}} \right]$$

A smaller unit of absolute viscosity is the centipoise, which is one-hundredth of a poise or:

$$1 \text{ centipoise} = 0.01 \text{ poise}$$

Kinematic Viscosity

The concept of kinematic viscosity is the outgrowth of the use of a head of liquid to produce a flow through a capillary tube. The coefficient of absolute viscosity, when divided by the density of the liquid, is called the kinematic viscosity. In the metric system, the unit of viscosity is called the *stoke* and it has the units of centimeters squared per second. One one-hundredth of a stoke is a centistoke.

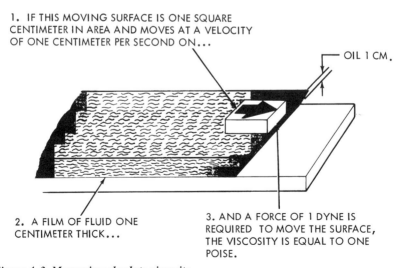

1. IF THIS MOVING SURFACE IS ONE SQUARE CENTIMETER IN AREA AND MOVES AT A VELOCITY OF ONE CENTIMETER PER SECOND ON...

OIL 1 CM.

2. A FILM OF FLUID ONE CENTIMETER THICK...

3. AND A FORCE OF 1 DYNE IS REQUIRED TO MOVE THE SURFACE, THE VISCOSITY IS EQUAL TO ONE POISE.

Figure 4–3 Measuring absolute viscosity.

The relationship between absolute and kinematic viscosity can be stated as:

$$\text{Centipoise} = \text{Centistoke} \times \text{Density or,}$$

$$\text{Centistroke} \ = \ \frac{\text{Centipoise}}{\text{Density}}$$

SUS Viscosity

For most practical purposes, it will serve to know the relative viscosity of the fluid. Relative viscosity is determined by timing the flow of a given quantity of the hydraulic fluid through a standard orifice at a given temperature. There are several methods in use. The most acceptable method in the United States is the *Saybolt viscosimeter* (Figure 4-4).

The time it takes for the measured quantity of liquid to flow through the orifice is measured with a stopwatch. The viscosity in *Saybolt Universal Seconds* (SUS) equals the elapsed time.

Obviously, a thick liquid will flow slowly, and the SUS viscosity will be higher than for a thin liquid, which flows faster. Since oil becomes thicker at low temperatures and thins when warmed, the viscosity must be expressed as a specific SUS at a given temperature. Tests are usually conducted at either 100°F or 210°F.

For industrial applications, hydraulic oil viscosity is typically approximately 150 SUS at 100°F. It is a general rule that the viscosity should never go below 45 SUS or above 4,000 SUS, regardless of temperature. Where temperature extremes are encountered, the fluid should have a high viscosity index.

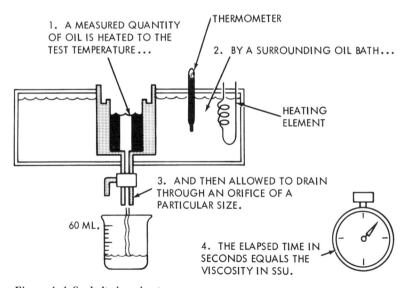

Figure 4–4 Saybolt viscosimeter.

SAE Number

SAE numbers have been established by the Society of Automotive Engineers to spec-ify ranges of SUS viscosities of oils at SAE test temperatures. Winter numbers (5W, 10W, 20W) are determined by tests at 0°F. Summer numbers (20W, 30W, etc.) desig-nate the SUS range at 210°F. Table 4-1 is a chart of the temperature ranges.

The following formulas may be used to convert centistokes (cSt units) to approximate Saybolt universal seconds (SUS units). For SUS values between 32 and 100:

$$cST = 0.226 \times SUS = \frac{193}{SUS}$$

Table 4–1 SAE Viscosity Numbers for Crankcase Oils

SAE Viscosity number	Viscosity units[a]	Viscosity range[b]			
		At 0°F		At 210°F	
		Minimum	Maximum	Minimum	Maximum
5W	Centipoise	—	Less than 1,200	—	—
	Centistokes	—	1,300	—	—
	SUS	—	6,000	—	—
10W	Centipoise	1,200[c]	Less than 2,400	—	—
	Centistokes	1,300	2,600	—	—
	SUS	6,000	12,000	—	—
20W	Centipoise	2,400[d]	Less than 9,600	—	—
	Centistokes	2,600	10,500	—	—
	SUS	12,000	48,000	—	—
20					Less than
	Centistokes	—	—	5.7	9.6
	SUS	—	—	45	58
30					Less than
	Centistokes	—	—	9.6	12.9
	SUS	—	—	58	70
40					Less than
	Centistokes	—	—	12.9	16.8
	SUS	—	—	70	85
50					Less than
	Centistokes	—	—	16.8	22.7
	SUS	—	—	85	110

a. The official values in this classification are based on 210°F viscosity in centistokes (ASTM D 445) and 0°F viscosities in centipoise (ASTM D260-2). Approximate values in other units of viscosity are given for information only. The approximate values at 0°F were calculated using an assumed oil density of 0.9 g/cm^3 at that temperature.

b. The viscosity of all oils included in this classification shall not be less than 3.0 centistokes at 210°F (39 SUS).

c. Minimum viscosity at 0°F may be waived provided viscosity at 210°F is not below 4.2 centistokes (40 SUS).

d. Minimum viscosity at 0°F may be waived provided viscosity at 210°F is not below 5.7 centistokes (45 SUS).

For SUS values greater than 100:

$$cST = 0.220 \times SUS = \frac{135}{SUS}$$

Although the viscometers just discussed are used in laboratories, there are other viscometers in the supply system that are available for local use. These viscometers can be used to test the viscosity of hydraulic fluids either prior to their being added to a system or periodically after they have been in an operating system for a while.

Viscosity Index

The viscosity index, VI, of oil is a number that indicates the effect of temperature changes on the viscosity of the oil. A low VI signifies a relatively large change of viscosity with changes of temperature. In other words, the oil becomes extremely thin at high temperatures and extremely thick at low temperatures. On the other hand, a high VI signifies relatively little change in viscosity over a wide temperature range. Figure 4-5 illustrates the relative change of viscosity with changes in oil temperature.

The ideal oil for most purposes is one that maintains a constant viscosity throughout temperature changes. The importance of the VI can be shown easily by considering automotive lubricants. An oil having a high VI resists excessive thickening when the engine is cold and, consequently, promotes rapid starting and prompt circulation; it resists excessive thinning when the motor is hot and thus provides full lubrication and prevents excessive oil consumption.

Another example of the importance of the VI is the need for a high-viscosity-index hydraulic oil for military aircraft, since hydraulic control systems may be exposed to

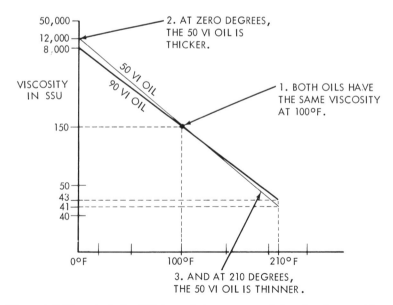

Figure 4–5 Viscosity index (VI) is a relative measure of viscosity change with temperature change.

temperatures ranging from below −65°F at high altitudes to over 100°F on the ground. For the proper operation of the hydraulic control system, the hydraulic fluid must have a sufficiently high VI to perform its functions at the extremes of the expected temperature range.

Liquids with a high viscosity have a greater resistance to heat than low-viscosity liquids, which have been derived from the same source. The average hydraulic liquid has a relatively low viscosity. Fortunately, there is a wide choice of liquids available for use in the viscosity range required of hydraulic liquids.

The VI of oil may be determined if its viscosity at any two temperatures is known. Tables, based on a large number of tests, are issued by the American Society for Testing and Materials (ASTM). These tables permit calculation of the VI from known viscosity.

Pour Point

Pour point is the lowest temperature at which a fluid will flow. If the hydraulic system will be exposed to extremely low temperature, it is a very important specification. As a rule of thumb, the pour point should be 20°F below the lowest temperature to be encountered.

Lubricating Power

If motion takes place between surfaces in contact, friction tends to oppose the motion. When pressure forces the liquid of a hydraulic system between the surfaces of moving parts, the liquid spreads out into a thin film that enables the parts to move more freely. Different liquids, including oils, vary greatly not only in their lubricating ability but also in film strength. Film strength is the capability of a liquid to resist being wiped or squeezed out from between the surfaces when spread out in an extremely thin layer. A liquid will no longer lubricate if the film breaks down, since the motion of part against part wipes the metal clean of liquid.

Lubricating power varies with temperature changes; therefore, the climatic and working conditions must enter into the determination of the lubricating qualities of a liquid. Unlike viscosity, which is a physical property, the lubricating power and film strength of a liquid are directly related to its chemical nature. Lubricating qualities and film strength can be improved by the addition of certain chemical agents.

Chemical Stability

Chemical stability is another property that is exceedingly important in the selection of a hydraulic liquid. It is defined as the liquid's ability to resist oxidation and deterioration for long periods. All liquids tend to undergo unfavorable changes under severe operating conditions. This is the case, for example, when a system operates for a considerable period of time at high temperatures.

Excessive temperatures, especially extremely high temperatures, have a great effect on the life of a liquid. The temperature of the liquid in the reservoir of an operating hydraulic system does not always indicate the operating conditions throughout the system. Localized hot spots occur on bearings, on gear teeth, or at other points where the liquid under pressure is forced through small orifices. Continuous passage of the liquid through these points may produce local temperatures high enough to carbonize the liquid or turn it into sludge, yet the liquid in the reservoir may not indicate an excessively high temperature.

Liquids may break down if exposed to air, water, salt, or other impurities, especially if they are in constant motion or subjected to heat. Some metals, such as zinc, lead, brass, and copper, have undesirable chemical reactions with certain liquids.

These chemical reactions result in the formation of sludge, gums, carbon, or other deposits that clog openings, cause valves and pistons to stick or leak, and give poor lubrication to moving parts. Once a small amount of sludge or other deposit has formed, the rate of formation generally increases more rapidly. As these deposits are formed, certain changes in the physical and chemical properties of the liquid take place. The liquid usually becomes darker, the viscosity increases, and damaging acids are formed.

The extent to which changes occur in different liquids depends on the type of liquid, type of refining, and whether it has been treated to provide further resistance to oxidation. The stability of liquids can be improved by the addition of oxidation inhibitors. Inhibitors selected to improve stability must be compatible with the other required properties of the liquid.

Freedom from Acidity

An ideal hydraulic liquid should be free from acids that cause corrosion of the metals in the system. Most liquids cannot be expected to remain completely noncorrosive under severe operating conditions. When new, the degree of acidity of a liquid may be satisfactory, but after use, the liquid may tend to become corrosive as it begins to deteriorate.

Many systems are idle for long periods after operating at high temperatures. This permits moisture to condense in the system, resulting in rust formation. Certain corrosion- and rust-preventive additives are added to hydraulic liquids. Some of these additives are effective only for a limited period. Therefore, the best procedure is to use the liquid specified for the system for the time specified by the system manufacturer and to protect the liquid and the system as much as possible from contamination by foreign matter, from abnormal temperatures, and from misuse.

Flashpoint

Flashpoint is the temperature at which a liquid gives off vapor in sufficient quantity to ignite momentarily or flash when a flame is applied. A high flashpoint is desirable for hydraulic liquids because it provides good resistance to combustion and a low degree of evaporation at normal temperatures. Required flashpoint minimums vary from 300°F for the lightest oils to 510°F for the heaviest oils.

Fire Point

Fire point is the temperature at which a substance gives off vapor in sufficient quantity to ignite and continue to burn when exposed to a spark or flame. Like flashpoint, a high fire point is required of desirable hydraulic liquids.

Minimum Toxicity

Toxicity is defined as the quality, state, or degree of being toxic or poisonous. Some liquids contain chemicals that are a serious toxic hazard. These toxic or poisonous chemicals may enter the body through inhalation, by absorption through the skin, or through the eyes or mouth. The result is sickness and, in some cases, death. Manufacturers of hydraulic liquids strive to produce suitable liquids that contain no toxic chemicals and, as a result, most hydraulic liquids are free of harmful chemicals. Some fire-resistant liquids are toxic, and suitable protection and care in handling must be provided.

Density and Compressibility

A fluid with a specific gravity of less than 1.0 is desired when weight is critical, although with proper system design, a fluid with a specific gravity greater than 1 can be tolerated. Where avoidance of detection by military units is desired, a fluid that sinks rather than rises to the surface of the water is desirable. Fluids having a specific gravity greater than 1.0 are desired, as leaking fluid will sink, allowing the vessel with the leak to remain undetected.

Under extreme pressure a fluid may be compressed up to 7 percent of its original volume. Highly compressible fluids produce sluggish system operation. This does not present a serious problem in small, low-speed operations, but it must be considered in the operating instructions.

Foaming Tendencies

Foam is an emulsion of gas bubbles in the fluid. In a hydraulic system, foam results from compressed gases in the hydraulic fluid. A fluid under high pressure can contain a large volume of air bubbles. When this fluid is depressurized, as when it reaches the reservoir, the gas bubbles in the fluid expand and produce foam. Any amount of foaming may cause pump cavitation and produce poor system response and sponge control. Therefore, defoaming agents are often added to fluids to prevent foaming. Minimizing air in fluid systems is discussed later in this chapter.

Cleanliness

Cleanliness in hydraulic systems has received considerable attention recently. Some hydraulic systems, such as aerospace hydraulic systems, are extremely sensitive to contamination. Fluid cleanliness is of primary importance because contaminants can cause component malfunction, prevent proper valve seating, cause wear in components, and may increase the response time of servo valves. Fluid contaminants are discussed later in this chapter.

The inside of a hydraulic system can only be kept as clean as the fluid added to it. Initial fluid cleanliness can be achieved by observing stringent cleanliness requirements (discussed later in this chapter) or by filtering all fluid added to the system.

TYPES OF HYDRAULIC FLUIDS

There have been many liquids tested for use in hydraulic systems. Currently, liquids being used include mineral oil, water, phosphate ester, water-based ethylene glycol compounds, and silicone fluids. The three most common types of hydraulic liquids are petroleum-based, synthetic fire-resistant, and water-based fire-resistant.

Petroleum-Based Fluids

The most common hydraulic fluids used in shipboard systems are the petroleum-based oils. These fluids contain additives to protect the fluid from oxidation (antioxidant), to protect system metals from corrosion (anti-corrosion), to reduce tendency of the fluid to foam (foam suppressant), and to improve viscosity.

Petroleum-based fluids are used in surface ships' electro-hydraulic steering and deck machinery systems, submarines' hydraulic systems, and aircraft automatic pilots, shock absorbers, brakes, control mechanisms, and other hydraulic systems using seal materials compatible with petroleum-based fluids.

Synthetic Fire-Resistant Fluids

Petroleum-based oils contain most of the desired properties of a hydraulic liquid. However, they are flammable under normal conditions and can become explosive when subjected to high pressures and a source of flame or high temperatures. Nonflammable synthetic liquids have been developed for use in hydraulic systems where fire hazards exist.

Phosphate Ester Fluid

As a maintenance person, operator, supervisor, or manager, you must understand the hazards associated with hydraulic fluids to which you may be exposed. This type of fluid contains a controlled amount of neurotoxic material. Because of the neurotoxic effects that can result from ingestion, skin absorption, or inhalation of these fluids, be sure to use the following precautions:

1. Avoid contact with the fluids by wearing protective clothing.
2. Use chemical goggles or face shields to protect your eyes.
3. If you are expected to work in an atmosphere containing a fine mist or spray, wear a continuous-flow airline respirator.
4. Thoroughly clean skin areas contaminated by this fluid with soap and water.
5. If you get any fluid in your eyes, flush them with running water for at least 15 minutes and seek medical attention.

If you come in contact with this type of fluid, report the contact when you seek medical aid and whenever you have a routine medical examination.

Silicone Synthetic Fire-Resistant Fluids

Silicone synthetic fire-resistant fluids are frequently used for hydraulic systems that require fire resistance, but that have only marginal requirements for other chemical or physical properties common to hydraulic fluids. Silicone fluids do not provide the corrosion protection and lubrication of phosphate ester fluids, but they also lack those fluids' detrimental characteristics, and they are excellent for fire protection.

Water-Based Fire-Resistant Fluids

The most widely used water-based hydraulic fluids may be classified as water–glycol mixtures and water–synthetic base mixtures. The water–glycol mixture contains additives to protect it from oxidation, corrosion, and biological growth and to enhance its load-carrying capacity.

Fire resistance of the water-mixture fluids depends on the vaporization and smothering effect of steam generated from the water. The water in water-based fluids is constantly being driven off while the system is operating. Therefore, frequent checks to maintain the correct ratio of water are important.

CONTAMINATION

Hydraulic fluid contamination may be described as any foreign material or substance whose presence in the fluid is capable of adversely affecting system performance or reliability. It may assume many different forms, including liquids, gases, and solid matter of various composition, sizes, and shapes. Solid matter is the type most often found in hydraulic systems and is generally referred to as particulate contamination. Contamination is always present to some degree, even in new, unused fluid, but must be kept below a level that will adversely affect system operation. Hydraulic contamination control consists of requirements, techniques, and practices necessary to minimize and control fluid contamination.

Classification of Contaminants

There are many types of contaminants, which are harmful to hydraulic systems and liquids. These contaminants may be divided into two different classes—particulate and fluid.

Particulate Contamination

This class of contaminants includes organic, metallic solid, and inorganic solid contaminants. These contaminants are discussed in the following paragraphs.

Organic. Wear, oxidation, or polymerization produces organic solids or semisolids found in hydraulic systems. Minute particles of O-rings, seals, gaskets, and hoses are present, due to wear or chemical reactions. Synthetic products, such as neoprene, silicones, and hypalon, though resistant to chemical reaction with hydraulic fluids, produce small wear particles. Oxidation of hydraulic fluids increases with pressure and temperature, although antioxidants are blended into hydraulic fluids to minimize such oxidation. The ability of a hydraulic fluid to resist oxidation or polymerization in service is defined as its oxidation stability. Oxidation products appear as organic acids, asphaltics, gums, and varnishes. These products combine with particles in the hydraulic fluid to form sludge. Some oxidation products are oil soluble and cause the hydraulic fluid to increase in viscosity; other oxidation products are not oil soluble and form sediment.

Metallic solids. Metallic contaminants are almost always present in a hydraulic system and will range in size from microscopic particles to particles readily visible to the naked eye. These particles are the result of wearing and scoring of bare metal parts and plating materials, such as silver and chromium. Although practically all metals commonly used for parts fabrication and plating may be found in hydraulic fluids, the major metallic materials found are ferrous, aluminum, and chromium particles. Because of their continuous high-speed internal movement, hydraulic pumps usually contribute most of the metallic particulate contamination present in hydraulic systems. Metal particles are also produced by other hydraulic system components, such as valves and actuators, due to body wear and the chipping and wearing away of small pieces of metal plating materials.

Inorganic solids. This contaminant group includes dust, paint particles, dirt, and silicates. Glass particles from glass bead peening and blasting may also be found as contaminants. Glass particles are very undesirable contaminants because of their abrasive effect on synthetic rubber seals and the very fine surfaces of critical moving parts. Atmospheric dust, dirt, paint particles, and other materials are often drawn into hydraulic systems from external sources. For example, the wet piston shaft of a hydraulic actuator may draw some of these foreign materials into the cylinder past the wiper and dynamic seals, and the contaminant materials are then dispersed in the hydraulic fluid. Contaminants may also enter the hydraulic fluid during maintenance when tubing, hoses, fittings, and components are disconnected or replaced. It is therefore important that all exposed fluid ports be sealed with approved protective closures to minimize such contamination.

Fluid Contamination

Air, water, solvent, and other foreign fluids are in the class of fluid contaminants.

Air. Hydraulic fluids are adversely affected by dissolved, entrained, or free air. Air may be introduced through improper maintenance or as a result of system design. Any maintenance operation that involves breaking into the hydraulic system, such as disconnecting or removing a line or component, will invariably result in some air being introduced into the system. This source of air can and must be minimized by prefilling replacement components with new filtered fluid prior to their installation. Failing to prefill a filter-element bowl with fluid is a good example of how air can be introduced into the system. Although prefilling will minimize introduction of air, it is still important to vent the system where venting is possible.

Most hydraulic systems have built-in sources of air. Leaky seals in gas-pressurized accumulators and reservoirs can feed gas into a system faster than it can be removed, even with the best of maintenance. Another lesser-known but major source is air that is sucked into the system past actuator piston rod seals. This occurs when the piston rod is stroked by some external means while the actuator itself is not pressurized.

Water. Water is a serious contaminant of hydraulic systems. Hydraulic fluids are adversely affected by dissolved, emulsified, or free water. Water contamination may result in the formation of ice, which impedes the operation of valves, actuators, and other moving parts. Water can also cause the formation of oxidation products and corrosion of metallic surfaces.

Solvent. Solvent contamination is a special form of foreign fluid contamination in which the original contaminating substance is a chlorinated solvent. Chlorinated solvents or their residues may, when introduced into a hydraulic system, react with any water present to form highly corrosive acids.

Chlorinated solvents, when allowed to combine with minute amounts of water often found in operating hydraulic systems, change chemically into hydrochloric acids. These acids then attack internal metallic surfaces in the system, particularly those that are ferrous, and produce a severe rustlike corrosion.

Foreign fluids. Foreign fluids other than water and chlorinated solvents can seriously contaminate hydraulic systems. This type of contamination is generally a result of lube oil, engine fuel, or incorrect hydraulic fluid being introduced inadvertently into the system during servicing. The effects of such contamination depend on the contaminant, the amount in the system, and how long it has been present.

Note: It is extremely important that the different types of hydraulic fluids not be mixed in one system. If different types are mixed, the characteristics of the fluid required for a specific purpose are lost. Mixing the different types of fluids usually will result in a heavy, gummy deposit that will clog passages and require a major cleaning. In addition, seals and packing installed for use with one fluid usually are not compatible with other fluids and damage to the seals will result.

Origin of Contamination

Recall that contaminants are produced from wear and chemical reactions, introduced by improper maintenance, and inadvertently introduced during servicing. These methods of contaminant introduction fall into one of the four major areas of contaminant origin.

1. *Particles originally contained in the system.* These particles originate during the fabrication and storage of system components. Weld spatter and slag may remain in welded system components, especially in reservoirs and pipe assemblies. The presence is minimized by proper design. For example, seam-welded overlapping joints are preferred, and arc welding of open sections is usually avoided. Hidden passages in valve bodies, inaccessible to sandblasting or other methods of cleaning, are the main sources of introduction of core sand. Even the most carefully designed and cleaned casting will almost invariably free some sand particles under the action of hydraulic pressure. Rubber hose assemblies always contain some loose particles. Most of these particles can be removed by flushing the hose before installation; however, some particles withstand cleaning and are freed later by the action of hydraulic pressure.

   Particles of lint from cleaning rags can cause abrasive damage in hydraulic systems, especially to closely fitted moving parts. In addition, lint in a hydraulic system packs easily into clearances between packing and contacting surfaces, leading to component leakage and decreased efficiency. Lint also helps clog filters prematurely. The use of the proper wiping materials will reduce or eliminate lint contamination. The wiping materials to be used for a given application will be determined by the following:

   (a) Substances being wiped or absorbed
   (b) The amount of absorbency required
   (c) The degree of cleanliness required

   These wiping materials are categorized for contamination control by the degree of lint or debris that they may deposit during use. For internal hydraulic repairs, this factor itself will determine the choice of wiping material.

   Rust or corrosion initially present in a hydraulic system can usually be traced to improper storage of materials and component parts. Particles can range in size from large flakes to abrasives of microscopic dimensions. Proper preservation of stored parts is helpful in eliminating corrosion.

2. *Particles introduced from outside sources.* Particles can be introduced into hydraulic systems at points where either the liquid or certain working parts of the system (for example, piston rods) are at least in temporary contact with the atmosphere. The most common contaminant introduction areas are at the refill and breather openings, cylinder rod packing, and open lines where components are removed for repair or replacement. Contamination arising from carelessness during servicing operations is minimized by the use of filters in the system fill lines and finger strainers in the filler adapter of hydraulic reservoirs. Hydraulic cylinder piston rods incorporate wiper rings and dust seals to prevent the dust that settles on the piston rod during

its outward stroke from entering the system when the piston rod retracts. Caps and plugs are available and should be used to seal off the open lines when a component is removed for repair or replacement.

3. *Particles created within the system during operation.* Contaminants created during system operation are of two general types—mechanical and chemical. Particles of a mechanical nature are formed by wearing of parts in frictional contact, such as pumps, cylinders, and packing gland components. These wear particles can vary from large chunks of packing down to steel shavings that are too small to be trapped by filters.

   The major source of chemical contaminants in hydraulic liquid is oxidation. These contaminants are formed under high pressures and temperatures and are promoted by the chemical action of water and air and of metals such as copper and iron oxides. Liquid-oxidation products appear initially as organic acids, asphaltines, gums, and varnishes—sometimes combined with dust particles as sludge. Liquid-soluble oxidation products tend to increase liquid viscosity, whereas insoluble types separate and form sediments, especially on colder elements such as heat exchanger coils.

   Liquids containing antioxidants have little tendency to form gums and sludge under normal operating conditions. However, as the temperature increases, resistance to oxidation diminishes. Hydraulic liquids that have been subjected to excessively high temperatures (above 250°F for most liquids) will break down, leaving minute particles of asphaltines suspended in the liquids. The liquid turns brown and is referred to as decomposed liquid. This explains the importance of keeping the hydraulic liquid temperature below specific levels.

   The second contaminant-producing chemical action in hydraulic liquids is one that permits these liquids to react with certain types of rubber. This reaction causes structural changes in the rubber, turning it brittle, and finally causing its complete disintegration. For this reason, the compatibility of system liquid with seals and hose material is a very important factor.

4. *Particles introduced by foreign liquids.* One of the most common foreign-fluid contaminants is water, especially in hydraulic systems that require petroleum-based liquids. Water, which enters even the most carefully designed system by condensation of atmospheric moisture, normally settles to the bottom of the reservoir. Oil movement in the reservoir disperses the water into fine droplets and agitation of the liquid in the pump and in high-speed passages forms an oil–water–air emulsion. This emulsion normally separates during the rest period in the system reservoir; but when fine dust and corrosion particles are present, high pressures chemically change the emulsion into sludge. The damaging action of sludge explains the need for effective filtration, as well as the need for water separation qualities in hydraulic liquids.

Control of Contamination

Maintaining hydraulic fluid within allowable contamination limits for both water and particulate matter is crucial to the care and protection of hydraulic equipment. Filters will provide adequate control of the particular contamination problem during all normal hydraulic system operations if the filtration system is installed properly and filter maintenance is performed regularly. Filter maintenance includes changing elements at proper intervals.

Control of the size and amount of contamination entering the system from any other source is the responsibility of the personnel who service and maintain the equipment. During installation, maintenance, and repair of hydraulic equipment, the retention of cleanliness of the system is of paramount importance for subsequent satisfactory performance.

The following maintenance and servicing procedures should be adhered to at all times to provide proper contamination control:

1. All tools and the work area (workbenches and test equipment) should be kept in a clean, dirt-free condition.
2. A suitable container should always be provided to receive the hydraulic liquid that is spilled during component removal or disassembly.

Note: The reuse of drained hydraulic liquid is prohibited in most hydraulic systems. In some large-capacity systems the reuse of fluid is permitted. When liquid is drained from these systems for reuse, it must be stored in a clean and suitable container. The liquid must be strained and/or filtered when it is returned to the system reservoir.

3. Before hydraulic lines or fittings are disconnected, the affected area should be cleaned with an approved dry-cleaning solvent.
4. All hydraulic lines and fittings should be capped or plugged immediately after disconnection.
5. Before any hydraulic components are assembled, their parts should be washed with an approved dry-cleaning solvent.
6. After the parts have been cleaned in dry-cleaning solvent, they should be dried thoroughly with clean, low-lint cloths and lubricated with the recommended preservative or hydraulic liquid before assembly.

Note: Only clean, low-lint type II or I cloth as appropriate should be used to wipe or dry component parts.

7. All packing and gaskets should be replaced during the assembly procedures.
8. All parts should be connected with care to avoid stripping metal slivers from threaded areas. All fittings and lines should be installed and torqued according to applicable technical instructions.
9. All hydraulic servicing equipment should be kept clean and in good operating condition.

Some hydraulic fluid specifications contain particle contamination limits that are so low that the products are packaged under clean-room conditions. Very slight amounts of dirt, rust, and metal particles will cause them to fail the specification limit for contamination. Since these fluids are usually all packaged in hermetically sealed containers, the act of opening a container may allow more contaminants into the fluid than the specification allows. Therefore, extreme care should be taken in the handling of these fluids. In opening the container for use, observation, or tests, it is extremely important that the can be opened and handled in a clean environment. The area of the container to be opened should be flushed with filtered solvent (petroleum ether or isopropyl alcohol), and the device used for opening the container should be thoroughly rinsed with filtered solvent. After the container is opened, a small amount of the material should be poured from the container and disposed of prior to pouring the sample for analysis. Once a container is opened, if the contents are not totally used, the unused portion should be discarded. Since the level of contamination of a system containing these fluids must be kept low, maintenance on the system's components must be performed in a clean environment commonly known as a controlled environment work center.

HYDRAULIC FLUID SAMPLING

The condition of a hydraulic system, as well as its probable future performance, can best be determined by analyzing the operating fluid. Of particular interest are any changes in the physical and chemical properties of the fluid and excessive particulate or water contamination, either of which indicates impending trouble.

Excessive particulate contamination of the fluid indicates that the filters are not keeping the system clean. This can result from improper filter maintenance, inadequate filters, or excessive ongoing corrosion and wear.

1. All samples should be taken from circulating systems, or immediately upon shutdown, while the hydraulic fluid is within 5°C (9°F) of normal system operating temperature. Systems not up to temperature may provide nonrepresentative samples of system dirt and water content, and such samples should either be avoided or so indicated on the analysis report. The first oil coming from the sampling point should be discarded, since it can be very dirty and does not represent the system. As a general rule, a volume of oil equivalent to one to two times the volume of oil contained in the sampling line and valve should be drained before the sample is taken.
2. Ideally, the sample should be taken from a valve installed specifically for sampling. When sampling valves are not installed, the taking of samples from locations where sediment or water can collect, such as dead ends of piping, tank drains, and low points of large pipes and filter bowls, should be avoided if possible. If samples are taken from pipe drains, sufficient fluid should be drained before the sample is taken to ensure that the sample actually represents the system. Samples are not to be taken from the tops

of reservoirs or other locations where the contamination levels are normally low.

3. Unless otherwise specified, a minimum of one sample should be taken for each system located wholly within one compartment. For ships' systems extending into two or more compartments, a second sample is required. An exception to this requirement is submarine external hydraulic systems, which require only one sample. Original sample points should be labeled and the same sample points used for successive sampling. If possible, the following sampling locations should be selected:

   (a) A location that provides a sample representative of fluid being supplied to system components
   (b) A return line as close to the supply tank as practical but upstream of any return line filter
   (c) For systems requiring a second sample, a location as far from the pump as practical

Operation of the sampling point should not introduce any significant amount of external contaminants into the collected fluid.

5

RESERVOIRS, STRAINERS, FILTERS, AND ACCUMULATORS

Fluid power systems must have a sufficient and continuous supply of uncontaminated fluid to operate efficiently. This chapter covers hydraulic reservoirs; various types of strainers and filters; and accumulators that are typically installed in fluid power systems.

RESERVOIRS

A hydraulic system must have a reserve of fluid in addition to that contained in the pumps, actuators, pipes, and other components of the system. This reserve fluid must be readily available to make up losses of fluid from the system, to make up for compression of fluid under pressure, and to compensate for the loss of volume as the fluid cools. This extra fluid is contained in a tank usually called a reservoir. A reservoir may sometimes be referred to as a sump tank, service tank, operating tank, supply tank, or base tank.

In addition to providing storage for the reserve fluid needed for the system, the reservoir acts as a radiator for dissipating heat from the fluid. It also acts as a settling tank where heavy particles of contamination may settle out of the fluid and remain harmlessly on the bottom until removed by cleaning or flushing the reservoir. Also, the reservoir allows entrained air to separate from the fluid.

Most reservoirs have a capped opening for filling, an air vent, an oil level indicator or dip stick, a return line connection, a pump inlet or suction line connection, a drain line connection, and a drain plug. See Figure 5-1.

The inside of the reservoir generally will have baffles to prevent excessive sloshing of the fluid and to put a partition between the fluid return line and the pump suction or inlet line. The partition forces the returning fluid to travel farther around the tank before being drawn back into the active system through the pump inlet line. This aids in settling the contamination and separating air entrained in the fluid.

67

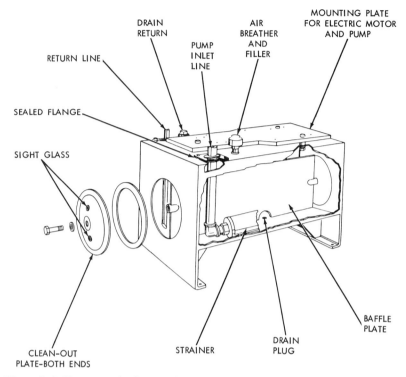

Figure 5–1 Nonpressurized reservoir.

Large reservoirs are desirable for cooling. A large reservoir also reduces recirculation, which helps settle contamination and separates entrained air. As a rule of thumb, the ideal reservoir should be two to three times the pump outlet per minute. However, because of space limitations in mobile and aerospace systems, the benefits of a large reservoir may have to be sacrificed. But, they must be large enough to accommodate thermal expansion of the fluid and changes in fluid level due to system operation.

ACCUMULATORS

An accumulator is a pressure storage reservoir in which hydraulic fluid is stored under pressure from an external source. The storage of fluid under pressure serves several purposes in hydraulic systems.

In some hydraulic systems, it is necessary to maintain the system pressure within a specific pressure range for long periods of time. It is very difficult to maintain a closed system without some leakage, either external or internal. Even a small leak can cause a decrease in pressure. By using an accumulator, leakage can be compensated for and the system pressure can be maintained within acceptable range for extended periods of time. Accumulators also compensate for thermal expansion and contraction of the liquid due to variations in temperature or generated heat.

A liquid flowing at a high velocity in a pipe will create a backward surge when stopped suddenly. This sudden stoppage causes an instantaneous pressure two to three times the operating pressure of the system. These pressures or shocks produce objectionable noise and vibrations, which can cause considerable damage to piping, fittings, and components. The incorporation of an accumulator enables such shocks and surges to be absorbed or cushioned by the entrapped gas, thereby reducing their effects. The accumulator also dampens pressure surges caused by pulsing delivery from the pump.

There are times when hydraulic systems require large volumes of liquid for short periods of time. This is due to either the operation of a large cylinder or the necessity of operating two or more circuits simultaneously. It is not economical to install a pump of such large capacity in the system for only intermittent usage, particularly if there is sufficient time during the working cycle for an accumulator to store enough liquid to aid the pump during these peak demands. The energy stored in accumulators may be also used to actuate hydraulically operated units if normal hydraulic system failure occurs.

Piston-Type Accumulators

Piston-type accumulators consist of a cylindrical body, called a barrel, closures on each end, called heads, and an internal piston. The piston may be fitted with a tailrod, which extends through one end of the cylinder (Figure 5-2), or it may not have a tailrod at all. In the latter case, it is referred to as a floating piston. Hydraulic fluid is pumped into one end of the cylinder, and the piston is forced toward the opposite end

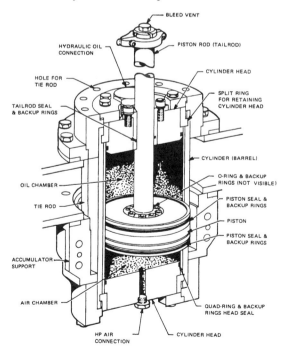

Figure 5–2 Cross-section of piston-type accumulator with a tailrod.

of the cylinder against a captive charge of air or an inert gas, such as nitrogen. Sometimes the amount of air charge is limited to the volume within the accumulator; other installations may use separate air flasks that are piped to the air side of the accumulator. Piston accumulators may be mounted in any position.

The gas portion of the accumulator may be located on either side of the piston. The orientation and type of accumulator are based on such criteria as available space, maintenance accessibility, size, need for external monitoring, contamination tolerance, seal life, and safety. The purpose of the piston seals is to keep the fluid and gas separate.

Usually, tailrod accumulators use two piston seals, one for the air side and one for the oil side, with the space between them vented to the atmosphere through a hole drilled the length of the tailrod. When the piston seals fail in this type of accumulator, air or oil leakage is apparent. However, seal failure in a floating piston or nonvented tailrod accumulators will not be as obvious. Therefore, more frequent attention to venting or draining the airside is necessary. An indication of worn and leaking seals can be detected by the presence of significant amounts of oil in the airside.

Bladder-Type Accumulators

Bladder- or bag-type accumulators consist of a shell or case with a flexible bladder inside the shell. See Figure 5-3.

The bladder is larger in diameter at the top (near the air valve) and gradually tapers to a smaller diameter at the bottom. The synthetic rubber is thinner at the top of the bladder than at the bottom. The operation of the accumulator is based on Barlow's formula for

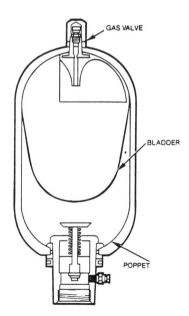

Figure 5–3 Bladder-type accumulator.

hoop stress, which states: "The stress in a circle is directly proportional to its diameter and wall thickness." This means that for a certain thickness, a large-diameter circle will stretch faster than a small-diameter circle; or for a certain diameter, a thin-wall hoop will stretch faster than a thick-wall hoop. Thus, the bladder will stretch around the top at its largest diameter and thinnest wall thickness, and then will gradually stretch downward and push itself outward against the walls of the shell. As a result, the bladder is capable of squeezing out all the liquid from the accumulator. Consequently, the bladder accumulator has a very high volumetric efficiency. In other words, this type of accumulator is capable of supplying a large percentage of the stored fluid to do work.

The bladder is precharged with air or inert gas to a specified pressure. Fluid is then forced into the area around the bladder, further compressing the gas in the bladder. This type of accumulator has the advantage that as long as the bladder is intact there is no exposure of fluid to the gas charge and therefore less danger of an explosion.

Diaphragm Accumulators

The diaphragm-type accumulator is constructed in two halves, which are either screwed or bolted together. A synthetic rubber diaphragm is installed between the two halves, making two separate chambers. Two threaded openings exist in the assembled component.

The opening at the top, as shown in Figure 5-4, contains a screen disc, which prevents the diaphragm from extruding through the threaded opening when system pressure is depleted, thus rupturing the diaphragm.

On some designs, a button-type protector fastened to the center of the diaphragm replaces the screen. An air valve for pressurizing the accumulator is located in the gas chamber end of the sphere, and the liquid port to the hydraulic system is located on the opposite end of the sphere. This accumulator operates in a manner similar to the bladder type.

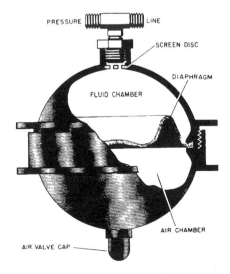

Figure 5–4 Diaphragm accumulator.

Accumulator Sizing

Most accumulator systems should be designed to operate at a maximum oil pressure of 3,000 psi. This is the rating of most accumulators and will give the maximum effect for the least cost. Also, 3,000 psi is the maximum rating for most hydraulic valves.

A rule of thumb for the nitrogen precharge level is one-half the maximum oil pressure. This is acceptable for most applications. The precharge should be replenished when it falls to one-third the maximum hydraulic oil pressure. On a 3,000-psi hydraulic system, initial precharge should be 1,500 psi and replenishment level of 1,000 psi. Most applications will tolerate a wide variation in precharge pressure.

Accumulators are catalog-rated by gas volume when all oil is discharged and usually rated in quarts or gallons (i.e., 1 U.S. gallon = 231 cubic inches). The amount of oil that can be stored is approximately half the gas volume. Only a part of the stored oil can be used each cycle because the oil pressure decreases as oil is discharged. The problem in selecting accumulator size is to have sufficient capacity that system pressure, at the end of the discharge, does not fall below a value that will do the job.

For illustration we are using an application in which accumulator oil will be used on the extension stroke of a cylinder to supplement the oil delivery from a pump, to increase speed. Retraction will be by pump volume alone. A fully charged accumulator system pressure of 3,000 psi is assumed.

First, select cylinder bore for sufficient force not only at 3,000 psi but also at some selected lower pressure to which it will be allowed to fall during discharge. Next, calculate the number of cubic inches of oil required to fill the cylinder cavity during its extension stroke. Using the time, in seconds, allowed for the full extension stroke, calculate the cubic inches of oil that can be obtained for the pump alone. Subtract the calculated pump volume from the cylinder volume to find the volume of oil required from the accumulator. Use Table 5-1 to find how many cubic inches of oil would be supplied from a 1-gallon accumulator before its terminal pressure dropped below the minimum acceptable pressure level. A 5-gallon accumulator would supply five times this volume. Finally, divide this figure into the total cubic inches needed for the application. This is the minimum rated gallon size of accumulator capacity. Select at least the next larger standard size for your application.

To solve for oil recovery from any size accumulator, under any system pressure and any precharge level, use the formula

$$D = [0.95 \times P_1 \times V_1 \div P_2] - [0.95 \times P_1 \times V_1 \div P_3],$$

where D is cubic inches of oil discharge; P_1 is precharge pressure in psi; P_2 is system pressure after volume D has been discharged; P_3 is maximum system pressure at full accumulator charge; V_1 is catalog-rated gas volume, in cubic inches; and 0.95 is assumed accumulator efficiency.

Table 5–1 Accumulator Selection Table

Minimum acceptable system psi	Cubic inch discharge
2,700	12
2,600	17
2,500	22
2,400	27
2,300	33
2,200	40
2,100	46
2,000	55
1,900	63
1,800	73
1,700	84
1,600	96
1,500	109

As oil is pumped into the accumulator, compressing the nitrogen, the nitrogen temperature increases (Charles' law). Therefore, the amount of oil stored will not be quite as much as calculated with Boyle's law unless sufficient time is allowed for the accumulator to cool to atmospheric temperature. Likewise, when oil is discharged, the expanding nitrogen is cooled. So the discharge volume will not be quite as high as calculated with Boyle's law. In the table and formula above, an allowance of 5 percent has been included as a safety factor. After making a size calculation from the table, allow enough extra capacity for contingencies.

FILTRATION

Clean hydraulic fluid is essential for proper operation and acceptable component life in all hydraulic systems. Although every effort must be made to prevent contaminants from entering the system, contaminants that do find their way in must be removed. Filtration devices are installed at key points in fluid power systems to remove the contaminants that enter the system along with those that are generated during normal operations of the system.

The filtering devices used in hydraulic systems are commonly referred to as strainers and filters. Since they share a common function, the terms are often used interchangeably. As a general rule, devices used to remove large particles of foreign matter from hydraulic systems are referred to as strainers, while those used to remove the smallest particles are called filters.

Strainers

Strainers are used primarily to catch only very large particles and will be found in applications where this type of protection is required. Most hydraulic systems have a strainer in the reservoir at the inlet to the suction line of the pump. A strainer is used in lieu of a filter to reduce its chance of being clogged and starving the pump. However, since this strainer is located in the reservoir, its maintenance is frequently neglected. When heavy dirt and sludge accumulate on the suction strainer, the pump soon begins to cavitate. Pump failure follows quickly.

Filters

The most common device installed in hydraulic systems to prevent foreign particles and contamination from remaining in the system are referred to as filters. They may be located in the reservoir, in the return line, in the pressure line, or in any other location in the system where the designer of the system decides they are needed to safeguard the system against impurities.

Filters are classified as full-flow or proportional flow. In full-flow types of filters, all of the fluid that enters the filter passes through the filtering element, whereas in proportional types only a portion of the fluid passes through the element.

Full-Flow Filters

The full-flow filter provides a positive filtering action. However, it offers resistance to flow, particularly when the filter element becomes dirty. Hydraulic fluid enters the filter through the inlet port in the body and flows around the filter element inside the filter bowl. Filtering takes place as the fluid passes through the filtering element and into the hollow core, leaving the dirt and impurities on the outside of the filter element.

The filtered fluid then flows from the hollow core through the outlet port and into the system. Figure 5-5 illustrates a typical full-flow filter. Some full-flow filters are equipped with a contamination indicator. These indicators, also known as differential pressure indicators, are available in three types: gauge, mechanical pop-up, and electrical. As contaminating particles collect on the filter element, the differential pressure across the element increases. In some installations using gauges as indicators, the differential pressure must be obtained by subtracting the readings of two gauges located somewhere along the filter inlet and outlet piping. For pop-up indicators, when the increase in pressure reaches a specific valve, an indicator (usually on the filter head) pops out, signifying that the filter must be cleaned or replaced. A low-temperature lockout feature is installed in most pop-up types of contamination indicators to eliminate the possibility of false indications due to cold weather.

Filter elements used in filters that have a contamination indicator are not normally removed or replaced until the indicator is actuated. This decreases the possibility of contaminated fluid bypassing the filter element and contaminating the entire system. This type of filter will minimize the necessity for flushing the entire system and lessen the possibility of failure of pumps and other components in the system.

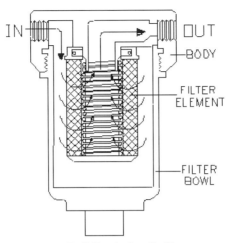

Figure 5–5 Full-flow hydraulic filter.

A bypass relief valve is installed in some filters. The bypass relief valve allows the fluid to bypass the filter element and pass directly through the outlet port in the event that the filter element becomes clogged. These filters may or may not be equipped with a contamination indicator. Figure 5-6 shows a full-flow, bypass-type hydraulic filter. This example includes a bypass indicator. Figure 5-7 shows a similar filter without the indicator.

A filter bypass indicator provides a positive indication, when activated, that the fluid is bypassing the filter element. This indicator should not be confused with the pop-up differential pressure indicator.

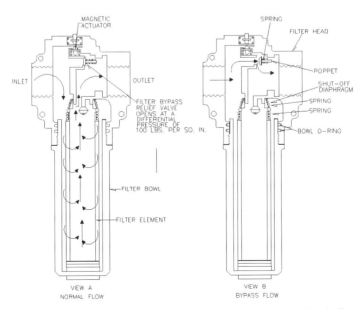

Figure 5–6 Full-flow bypass-type filter (with contamination indicator).

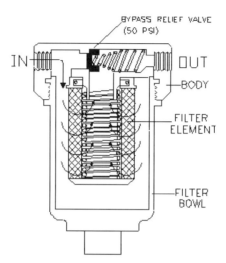

Figure 5–7 Full-flow bypass hydraulic filter.

Proportional-Flow Filters

This type of filter operates on the venturi principle. As the fluid passes through the venturi throat, a drop in pressure is created at the narrowest point. Figure 5-8 illustrates a proportional-flow filter. A portion of the fluid flowing toward and away from the throat of the venturi flows through this passage into the body of the filter. A fluid passage connects the hollow core of the filter with the throat of the venturi. Thus, the low-pressure area at the throat of the venturi causes the pressurized fluid in the body of the filter to flow through the filter element, through the hollow core, and into the low-pressure area. The fluid is then returned to the system. Although only a portion of

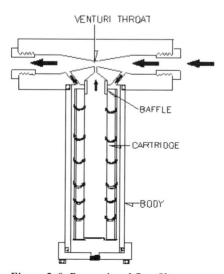

Figure 5–8 Proportional flow filter.

the fluid is filtered during each cycle, constant recirculation through the system will eventually cause all the fluid to pass through the filter element.

Filter Rating

Filters are rated in several ways: absolute, mean, and nominal. The absolute filtration rating is the diameter, in microns, of the largest spherical particle that will pass through the filter under a certain test condition. This rating is an indication of the largest opening in the filter element. The mean filtration rating is the measurement of the average size of the openings in the filter element. The nominal filtration rating is usually interpreted to mean the size of the smallest particles of which 90 percent will be trapped in the filter at each pass through the filter element.

Filter Elements

Filter elements may be divided into two classes: surface and depth. Surface filters are made of closely woven fabric or treated paper with a uniform pore size. Fluid flows through the pores of the filter material and contaminants are stopped on the filter's surface. This type of filter element is designed to prevent the passage of a high percentage of solids of a specific size.

Depth filters, on the other hand, are composed of layers of fabric or fibers, which provide many tortuous paths for the fluid to flow through. The pores or passages must be larger than the rated size of the filter if particles are to be retained in the depth of the medium rather than on the surface.

Filter elements may be of the 5-micron, woven-mesh, micronic, porous-metal, or magnetic type. The micronic and 5-micron elements have noncleanable filter media and are disposed of when they are removed. Porous-metal, woven-mesh, and magnetic filter elements are usually designed to be cleaned and reused.

Noncleanable 5-micron. The most common 5-micron filter medium is composed of organic and inorganic fibers integrally bonded by epoxy resin and faced with a metallic mesh upstream and downstream for protection and added mechanical strength. Filters of this type are not to be cleaned under any circumstances.

Another 5-micron filter medium uses layers of very fine stainless-steel fibers drawn into a random but controlled matrix. Filter elements of this material may be either cleanable or disposable, depending on their construction.

Woven wire mesh. Filters of this type are made of stainless steel and are generally rated as 15 or 25 microns (absolute). This type of element is reusable.

Micronic hydraulic. The term micronic is derived from the word micron. It could be used to describe any filter element; however, through usage this term has become associated with a specific filter with a specially treated cellulose paper (Figure 5-9).

Figure 5–9 Micronic filter element.

Magnetic filters. Some hydraulic systems have magnetic filters installed at strategic points. Filters of this type are designed primarily to trap any ferrous particles that may be in the system.

HEAT EXCHANGERS

The conversion of hydraulic force to mechanical work generates excessive heat. This heat must be removed from the hydraulic fluid to prevent degradation of the fluid and possible damage to system components.

Heat goes into the hydraulic oil at every point in the system where there is a pressure loss due to oil flow without mechanical work being produced. Examples are pressure relief and reducing valves, flow control valves, and flow resistance in plumbing lines and through components. Hydraulic pumps and motors also produce heat at about 15 percent of their working horsepower. Power loss and heat generation due to these causes can be calculated with one of the following formulas:

$$\text{Horsepower Heat} = \frac{\text{PSI} \times \text{GPM}}{1,714} \text{ or HP} = \text{PSI} \times \text{GPM} \times .000583$$

$$\text{BTU/Hr of heat generation} = 1.5 \times \text{PSI} \times \text{GPM}$$

The pressure (psi) for calculating heat generation in a flow control valve, for example, is the inlet minus the outlet pressure, or the pressure drop across the valve.

Sometimes power loss and heat generation occur intermittently, and to find the average amount of heat that will go into the oil, the *average* power loss should be calculated. Usually, taking an average over a 1-hour period should be sufficient. Therefore, hydraulic systems should include a positive means of heat removal. Normally, a heat exchanger is used for this purpose.

The exact size of a heat exchanger needed on a new system cannot be accurately calculated because of too many unknown factors. On existing systems, by making tank measurements and measuring air and oil temperatures, a rather accurate calculation can be made of the heat exchanger capacity needed to reduce the maximum oil temperature.

The theoretical maximum cooling capacity of a heat exchanger for a hydraulic system will never have to be greater than the input horsepower to the system. Usually its capacity can be considerably less, based on the calculated input horsepower. A rule-of-thumb is to provide a heat exchanger removal capacity of about 25 percent of the input horsepower. Rarely, even on inefficient systems, would a capacity of more than 50 percent be required.

In ordering a heat exchanger, the information furnished to the supplier must include the maximum rate of oil flow, in GPM, through the heat exchanger, and the horsepower or BTU per hour of heat to be removed. On water-cooled models, state the maximum rate of water flow that will be available. For best water usage, the water flow should be approximately one-half of the oil flow. Specify the temperature of the cooling water.

Heat Load

Heat load on the water side of a shell and tube heat exchanger can be calculated as:

$$\text{BTU per Hour} = \text{GPM} \times 500 \times \text{Temperature Differential}$$

The temperature differential is the difference between the inlet and outlet oil temperature in degrees Fahrenheit.

Heat load on the shell side of the exchanger can also be calculated by:

$$\text{BTU per Hour} = \text{GPM} \times 210 \times \text{Temperature Differential}$$

Heat and Power Equivalents	
1 horsepower (HP)	2,545 BTU per hour
1 horsepower (HP)	42.4 BTU per minute
1 British thermal unit per hour (BTU/hour)	.000393 HP or 0.293 watts
1 British thermal unit per minute (BTU/min)	.0167 BTU/hour or 17.6 watts

6

ACTUATORS

One of the outstanding features of fluid power systems is that force, generated by the power supply, controlled and directed by suitable valving, and transported by lines, can be converted with ease to almost any kind of mechanical motion. Either linear or rotary motion can be obtained by using a suitable actuating device.

An actuator is a device that converts fluid power into mechanical force and motion. Cylinders, hydraulic motors, and turbines are the most common types of actuating devices used in fluid power systems. This chapter describes various types of actuating devices and their applications.

HYDRAULIC CYLINDERS

An actuating cylinder is a device that converts fluid power into linear, or straight-line, force and motion. Since linear motion is back-and-forth motion along a straight line, this type of actuator is sometimes referred to as a reciprocating or linear motor. The cylinder consists of a ram, or piston, operating within a cylindrical bore. Actuating cylinders may be installed so that the cylinder is anchored to a stationary structure and the ram or piston is attached to the mechanism to be operated, or the piston can be anchored and the cylinder attached to the movable mechanism.

Actuating cylinders for pneumatic and hydraulic systems are similar in design and operation. Some variations of ram- and piston-type actuating cylinders are described in the following pages.

Ram-Type Cylinders

The terms ram and piston are often used interchangeably. However, a ram-type cylinder is usually considered one in which the cross-sectional area of the piston rod is

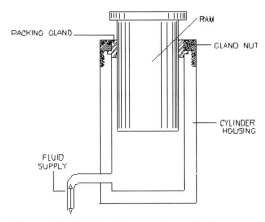

Figure 6–1 Single-acting ram-type cylinder.

more than one-half the cross-sectional area of the movable element. In most actuating cylinders of this type, the rod and the movable element have equal areas. This type of movable element is frequently referred to as a plunger.

The ram-type actuator is used primarily to push rather than to pull. Some applications require simply a flat surface on the external part of the ram for pushing or lifting the unit to be operated. Other applications require some mechanical means of attachment, such as a clevis or eyebolt. The design of ram-type cylinders varies in many other respects to satisfy the requirements of different applications.

Single-Acting Ram

The single-acting ram (Figure 6-1) applies force in only one direction. The fluid that is directed into the cylinder displaces the ram and forces it outward, lifting the object placed on it.

Since there is no provision for retracting the ram by fluid power, when fluid pressure is released, either the weight of the object or some mechanical means, such as a spring, forces the ram back into the cylinder. This forces the fluid back to the reservoir.

Double-Acting Ram

A double-acting ram-type cylinder is illustrated in Figure 6-2. In this cylinder, both strokes of the ram are produced by pressurized fluid. There are two fluid ports, one at each end of the cylinder tube. Fluid, under pressure, is directed to the closed end of the cylinder to extend the ram and apply force. To retract the ram and reduce the force, fluid is directed to the opposite end of the cylinder.

A four-way directional control valve is normally used to control the double-acting ram. When the valve is positioned to extend the ram, pressurized fluid enters port A (Figure 6-2), acts on the bottom surface of the ram, and forces the ram outward. Fluid

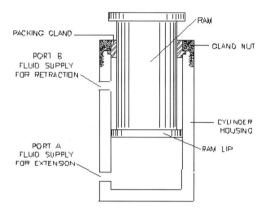

Figure 6–2 Double-acting ram-type cylinder.

above the ram lip is free to flow out of port B, through the control valve, and back to the reservoir.

Normally, the pressure of the fluid is the same for either stroke of the ram. Recall from earlier discussions that force is equal to pressure times area ($F = PA$). Notice the difference of the areas upon which pressure acts in Figure 6-2. The pressure acts against the large surface area on the bottom of the ram during the extension or pressure stroke. Since the ram does not require a large force during the retraction stroke, pressure acting on the small area on the top surface of the ram lip provides the necessary force to retract the ram.

Telescoping Rams

Figure 6-3 shows a telescoping ram-type actuating cylinder. A series of rams is nested in the smallest ram; each ram is hollow and serves as the cylinder housing for the next smaller ram. The ram assembly is contained in the main cylinder assembly, which also provides the fluid ports. Although the assembly requires a small space with all the rams retracted, the telescoping action of the assembly provides a relatively long stroke when the rams are fully extended.

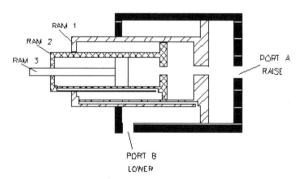

Figure 6–3 Telescoping ram-type cylinder.

An excellent example of the application of this type of cylinder is the dump truck. The cylinder is used to lift the forward end of the truck bed and dump the load. During the lifting operation, the greatest force is required for the initial lift. As the load is lifted and begins to dump, the required force becomes less and less until the load is completely dumped. During the raise cycle, pressurized fluid enters the cylinder through port A (Figure 6-3) and acts on the bottom surface of all three rams. Ram 1 has the larger surface area and therefore provides the greater force. As ram 1 reaches the end of its stroke the required force is decreased, and ram 2 moves, raising the load. When ram 2 completes its stroke, a still smaller force is required. Ram 3 then moves outward to finish raising and dumping the load.

Piston-Type Cylinders

An actuating cylinder in which the cross-sectional area of the piston is less than one-half the cross-sectional area of the movable element is referred to as a piston-type cylinder. This type of cylinder is normally used for applications that require both push and pull functions. The piston-type cylinder is the most common type used in fluid power systems.

The essential parts of a piston-type cylinder are a cylindrical barrel, a piston and rod, end caps, and suitable seals. The end caps are attached to the ends of the barrel. These end caps usually contain the fluid ports. The end cap on the rod end contains a hole for the piston rod to pass through. Suitable seals are used between the hold and the piston rod to keep fluid from leaking out and to keep dirt and other contamination from entering the barrel. The opposite end cap of most cylinders is provided with a fitting for securing the actuating cylinder to some structure. This end cap is referred to as the anchor end cap.

The piston rod may extend through either or both ends of the cylinder. The extended end of the rod is normally threaded so that some type of mechanical connection, such as an eyebolt or clevis, can be attached. This threaded connection provides for adjustment between the rod and the unit to be attached. After the correct adjustment is made, the locknut is tightened against the connector to prevent the connector from turning. The other end of the connector is attached, either directly or through additional mechanical linkage, to the unit to be actuated.

In order to satisfy the many requirements of fluid power systems, piston-type cylinders are available in a variety of designs.

Single-Acting Cylinder

The single-acting cylinder is similar in design and operation to the single-acting ram-type cylinder. The single-acting piston-type cylinder uses fluid pressure to provide the force to extend or retract the cylinder and spring tension, gravity, or some other outside force to move the cylinder in the opposite direction. Figure 6-4 shows a single-acting spring-loaded cylinder. In this cylinder, the spring is located on the rod side of

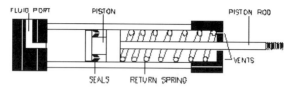

Figure 6–4 Single-acting spring-loaded cylinder.

the piston. In some spring-loaded cylinders the spring is located on the blank side and the fluid port is on the rod side of the cylinder.

A three-way directional control valve is normally used to control this type of cylinder. To extend the piston rod, pressurized fluid is directed through the port into the cylinder (Figure 6-4). This pressure acts on the surface area of the blank side of the piston and forces the piston to the right. This action moves the rod to the right, through the end of the cylinder, thus moving the actuated unit in one direction. During this action, the spring is compressed between the rod side of the piston and the end of the cylinder. The length of the stroke depends upon the physical limits within the cylinder and the required movement of the actuated unit.

To retract the piston rod, the directional control valve is moved to the opposite working position, which releases the pressure in the cylinder. The spring tension forces the piston to the left, retracting the piston rod and moving the actuated unit in the opposite direction. The fluid is free to flow from the cylinder through the port, through the control valve, and to the reservoir.

The end of the cylinder opposite the fluid port is vented to atmosphere. This prevents air from being trapped in this area. Any trapped air would compress during the extension stroke, creating excess pressure on the rod side of the piston. This would cause sluggish movement of the piston and could eventually cause a complete lock, preventing the fluid pressure from moving the piston. Adequate seals prevent leakage between the cylinder wall and piston. The piston in Figure 6-4 contains V-ring seals that prevent blow-by of hydraulic fluid that would then be vented through the bleed ports.

Double-Acting Cylinders

Most piston-type cylinders are double acting, which means that pressurized fluid can be applied to either side of the piston. In this type of cylinder work is performed in two directions.

One design of the double-acting cylinder is shown in Figure 6-5. This cylinder contains one piston and piston rod assembly. The stroke of the piston and piston rod assembly in either direction is produced by fluid pressure. The two fluid ports, one near each end of the cylinder, alternate as inlet and outlet ports, depending on the direction of flow from the directional control valve. This actuator is referred to as an unbalanced actuator because there is a difference in the effective working areas on the two sides of the piston. Therefore, this type of cylinder is normally installed so that the blank side of the piston carries the greatest load—that is, the cylinder carries the greater load during the extension stroke.

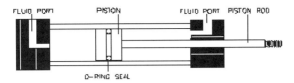

Figure 6–5 Double-acting piston-type cylinder.

A four-way directional control valve is normally used to control the operation of this type of cylinder. The valve can be positioned to direct pressurized fluid to either end of the cylinder and allow the displaced fluid to flow from the opposite end back to the reservoir.

There are applications where it is necessary to move two mechanisms at the same time. In this case, double-acting cylinders of different designs are required. Figures 6-6 and 6-7 illustrate examples of these applications.

Figure 6-6 shows a three-port double-acting cylinder. This actuator contains two pistons and piston rod assemblies. Fluid is directed through port A by a four-way valve and moves the pistons outward, thus moving the mechanisms attached to the piston rods. The fluid on the rod side of both pistons is forced out of the cylinder through ports B and C, which are connected by a common line to the directional control valve. The displaced fluid then flows through the control valve to the reservoir.

When pressurized fluid is directed into the cylinder through ports B and C, the two pistons move inward, also moving the mechanisms attached to both rods. Fluid between the two pistons and the center end cap is free to flow from the cylinder through port A, through the control valve, and back to the reservoir.

The double-acting cylinder in Figure 6-7 is a balanced actuator. The piston rod extends through the piston and out through both ends of the cylinder. Since both piston rods are the same diameter, the force of the cross-sectional area on both sides of the piston generated is equal in both directions. One or both ends of the piston rod may be attached to a mechanism to be operated.

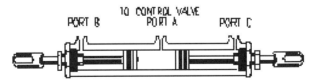

Figure 6–6 Three-ported double-acting cylinder.

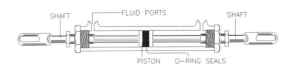

Figure 6–7 Balanced double-acting cylinder.

PISTON ROD COLUMN STRENGTH

Long, slim piston rods may buckle if subjected to too heavy a push load. Table 6-1 suggests the minimum diameter piston rod to use under various conditions of load and unsupported rod length. It should be used in accordance with the instructions in the following paragraph. There must be no side load or bending stress at any point along the piston rod.

Exposed rod length is shown along the top of the table. This is usually somewhat longer than the actual stroke of the cylinder. The vertical scale (column 1) shows the load on the cylinder and is expressed in English tons, i.e., 1 ton equals 2,000 pounds. If both the end of the rod and the *front* end of the cylinder barrel are rigidly supported, a smaller rod may have sufficient column strength, and you may use an *exposed* length of piston rod that is one-half of the actual total rod length. For example, if the actual rod length is 80 inches, and if the cylinder barrel and rod end are supported as described, you could enter the table in the column marked 40. On the other hand, if hinge mounting is used on both cylinder and rod, you may not be safe in using actual exposed rod length. Instead, you should use about twice the actual rod length. For example, if the actual rod length is 20 inches, you should enter the table in the 40-inch column.

When mounted horizontally or at any angle other than vertical, hinge-mounted cylinders create a bending stress on the rod when extended. In part, this bending stress is created by the cylinder's weight. On large-bore and/or long-stroke hinge-mounted cylinders, a trunnion mount should be used in instead of tang or clevis mounts. In addition, the trunnion should be positioned so that the cylinder's weight is balance when the rod is fully extended.

Table 6–1 Minimum Recommended Piston Rod Diameter

Load in tons	Exposed length of piston rod (inches)							
	10	20	40	60	70	80	100	120
1/2			3/4	1				
3/4			13/16	1-1/16				
1		5/8	7/8	1-1/8	1-1/4	1-3/8		
1-1/2		11/16	15/16	1-3/16	1-3/8	1-1/2		
2		3/4	1	1-3/16	1-7/16	1-9/16	1-13/16	
3	13/16	7/8	1-18	1-3/8	1-9/16	1-5/8	1-7/8	
4	15/16	1	1-3/16	1-1/2	1-5/8	1-3/4	2	2-1/4
5	1	1-1/8	1-5/16	1-9/16	1-3/4	1-7/8	2-1/8	2-3/8
7-1/2	1-3/16	1-1/4	1-7/16	1-3/4	1-7/8	2	2-1/4	2-1/2
10	1-3/8	1-7/16	1-5/8	1-7/8	2	2-1/8	2-7/16	2-3/4
15	1-11/16	1-3/4	1-7/8	2-1/8	2-1/4	2-3/8	2-11/16	3
20	2	2	2-1/8	2-3/8	2-1/2	2-5/8	2-7/8	3-1/4
30	2-3/8	2-7/16	2-1/2	2-3/4	2-3/4	2-7/8	3-1/4	3-1/2
40	2-3/4	2-3/4	2-7/8	3	3	3-1/4	3-1/2	3-3/4
50	3-1/8	3-1/8	3-1/4	3-3/8	3-1/2	3-1/2	3-3/4	4
75	3-3/4	3-3/4	3-7/8	4	4	4-1/8	4-3/8	4-1/2
100	4-3/8	4-3/8	4-3/8	4-1/2	4-3/4	4-3/4	4-7/8	5
150	5-3/8	5-3/8	5-3/8	5-1/2	5-1/2	5-1/2	5-3/4	6

HYDRAULIC CYLINDER FORCES

Tables 6-2 and 6-3 provide the mechanical forces, both extension and retraction, that can be generated by hydraulic cylinders. The tables are divided into the two principal operating pressure ranges associated with hydraulic applications.

Values in bold type show the extension forces, using the full piston area. Values in italic type are for the retraction force for various piston rod diameters. Remember that force values are *theoretical*, derived by calculation. Experience has shown that probably 5 percent but certainly no more than 10 percent additional pressure will be required to make up cylinder losses.

For pressures not shown, the effective piston areas in the third column can be used as power factors. Multiply effective area times pressure to obtain cylinder force produced. Values in two or more columns can be added for a pressure not listed, or force values can be obtained by interpolating between the next higher and the next lower pressure columns.

Pressure values along the top of each table are differential pressures across the two cylinder ports. This is the pressure to just balance the load and not the pressure that must be produced by the system pump. There will be circuit flow losses in pressure and return lines due to oil flow, and these will require additional pressure. When designing a system, be sure to allow sufficient pump pressure, about 25 to 30 percent, both to supply the cylinder and to satisfy system flow losses.

CYLINDER MOUNTING

Various cylinder mountings (Figure 6-8) provide flexibility in anchoring the cylinder. Rod ends are usually threaded for attachment directly to the load or to accept a clevis, yoke, or similar coupling device.

Foot and Centerline Lug Mounts

This type of mounting provides rigid attachment of the cylinder to a stationary, structural member of a machine. It also provides a firm foundation that equally transmits the forces generated by the hydraulic system, through the cylinder, to the movable machine component. Because of the rigid mount, this configuration provides the best transmission of force without induced sideload on the cylinder rod.

Front Flange Mount

As in the lug mount, the flange-mounted unit provides reasonably rigid attachment of the cylinder to a structural component of the machine. Because of the relatively small mounting area and absence of a rear mount, this configuration does not provide the full support of the lug mount design. As a result, care must be taken to minimize sideload or overextension of the cylinder rod.

Table 6–2 Hydraulic Cylinder Force: Low Pressure Range, 500 to 1,500 psi

Bore diameter (inches)	Rod diameter (inches)	Effective area (sq. in.)	Pressure differential across cylinder ports				
			500 PSI	750 PSI	1000 PSI	1250 PSI	1500 PSI
1-1/2	None	1.77	884	1,325	1,767	2,209	2,651
	5/8	1.46	730	1,095	1,460	1,825	2,190
	1	.982	491	736	982	1,227	1,473
2	None	3.14	1,571	2,356	3,142	3,927	4,712
	1	2.36	1,178	1,767	2,356	2,945	3,534
	1-3/8	1.66	828	1,243	1,657	2,071	2,485
2-1/2	None	4.91	2,454	3,682	4,909	6,136	7,363
	1	4.12	2,062	3,092	4,123	5,154	6,188
	1-3/8	3.42	1,712	2,568	3,424	4,280	5,136
	1-3/4	2.50	1,252	1,878	2,503	3,129	3,755
3	None	7.07	3,534	5,301	7,069	8,836	10,603
	1	6.28	3,142	4,712	6,283	7,854	9,425
	1-3/8	5.58	2,792	4,188	5,584	6,980	8,376
	1-3/4	4.66	2,332	3,497	4,663	5,829	6,995
3-1/4	None	8.30	4,148	6,222	8,298	10,370	12,444
	1-3/8	6.81	3,405	5,108	6,811	8,514	10,216
	1-3/4	5.89	2,945	4,418	5,891	7,363	8,836
	2	5.15	2,577	3,866	5,154	6,443	7,731
4	None	12.57	6,284	9,425	12,567	15,709	18,851
	1-3/4	10.16	5,081	7,621	10,162	12,702	15,243
	2	9.43	4,713	7,069	9,425	11,782	14,138
	2-1/2	7.66	3,829	5,744	7,658	9,573	11,487
5	None	19.64	9,818	14,726	19,635	24,544	29,453
	2	16.49	8,247	12,370	16,493	20,617	24,740
	2-1/2	14.73	7,363	11,045	14,726	18,408	22,089
	3	12.57	6,283	9,425	12,566	15,708	18,850
	3-1/2	10.01	5,007	7,510	10,014	12,517	15,021

Table 6–2 Continued.

Bore diameter (inches)	Rod diameter (inches)	Effective area (sq. in.)	Pressure differential across cylinder ports				
			500 PSI	750 PSI	1000 PSI	1250 PSI	1500 PSI
6	None	28.27	14,137	21,206	28,274	35,343	42,411
	2-1/2	23.37	11,683	17,524	23,365	29,207	35,048
	3	21.21	10,603	15,904	21,205	26,507	31,808
	3-1/2	18.65	9,326	13,990	18,653	23,316	27,979
	4	15.71	7,854	11,781	15,708	19,635	23,562
7	None	38.49	19,243	28,864	38,485	48,106	57,728
	3	31.42	15,708	23,562	31,416	39,271	47,125
	3-1/2	28.87	14,432	21,648	28,864	36,080	43,296
	4	25.92	12,960	19,439	25,910	32,399	38,879
	4-1/2	22.58	11,291	16,936	22,581	28,226	33,872
	5	18.85	9,425	14,138	18,850	23,563	28,275
8	None	50.27	25,133	37,699	50,265	62,831	75,398
	3-1/2	40.64	20,322	30,483	40,644	50,805	60,966
	4	37.70	18,850	28,274	37,699	47,124	56,549
	4-1/2	34.36	17,181	25,771	34,361	42,951	51,542
	5	30.63	15,315	22,973	30,630	38,288	44,945
	5-1/2	26.51	13,254	19,880	26,507	33,134	39,761
10	None	78.54	39,270	58,905	78,540	98,175	117,810
	4-1/2	62.64	31,318	46,977	62,636	78,295	93,954
	5	58.91	29,453	44,179	58,905	73,631	88,358
	5-1/2	54.78	27,391	41,087	54,782	68,478	82,172
	7	40.06	20,028	30,041	40,055	50,069	60,083
12	None	113.1	56,550	84,825	113,100	141,375	169,650
	5-1/2	89.34	44,671	67,007	89,342	111,678	134,013
14	None	153.9	76,970	115,455	153,940	192,425	230,910
	7	115.5	57,728	86,591	115,455	144,319	173,183

Table 6–3 Hydraulic Cylinder Force: High Pressure Range, 2,000 to 5,000 psi

Bore diameter (inches)	Rod diameter (inches)	Effective area (sq. in.)	Pressure differential across cylinder ports				
			2,000 PSI	2,500 PSI	3,000 PSI	4,000 PSI	5,000 PSI
1-1/2	None	**1.77**	**3534**	**4418**	**5301**	**7068**	**8836**
	5/8	1.46	2921	3651	4381	5841	7302
	1	.982	1963	2454	2945	3927	4909
2	None	**3.14**	**6283**	**7854**	**9425**	**12,566**	**15,708**
	1	2.36	4712	5890	7069	9425	11,781
	1-3/8	1.66	3313	4142	4970	6627	8283
2-1/2	None	**4.91**	**9817**	**12,271**	**14,726**	**19,635**	**24,544**
	1	4.12	8247	10,308	12,370	16,493	20,617
	1-3/8	3.42	6848	8560	10,271	13,695	17,119
	1-3/4	2.50	5007	6259	7510	10,014	12,517
3	None	**7.07**	**14,137**	**17,672**	**21,206**	**28,274**	**35,343**
	1	6.28	12,567	15,708	18,850	25,133	31,416
	1-3/8	5.58	11,167	13,959	16,751	22,335	27,919
	1-3/4	4.66	9327	11,658	13,990	18,653	23,317
3-1/4	None	**8.30**	**16,592**	**20,740**	**24,837**	**33,183**	**41,479**
	1-3/8	6.81	13,622	17,027	20,433	27,244	34,055
	1-3/4	5.89	11,781	14,726	17,672	23,562	29,453
	2	5.15	10,308	12,886	15,463	20,617	25,771
4	None	**12.57**	**25,134**	**31,418**	**37,701**	**50,268**	**62,835**
	1-3/4	10.16	20,323	25,404	30,485	40,647	50,809
	2	9.43	18,851	23,564	28,276	37,702	47,127
	2-1/2	7.66	15,317	19,146	22,975	30,633	38,292
5	None	**19.64**	**39,270**	**49,088**	**58,905**	**78,540**	**98,175**
	2	16.49	32,987	41,234	49,480	65,974	82,467
	2-1/2	14.73	29,453	36,816	44,179	58,905	73,632
	3	12.57	25,133	31,416	37,699	50,266	62,832
	3-1/2	10.01	20,028	25,035	30,042	40,056	50,070

Table 6–3 Continued.

Bore diameter (inches)	Rod diameter (inches)	Effective area (sq. in.)	Pressure differential across cylinder ports				
			2,000 PSI	2,500 PSI	3,000 PSI	4,000 PSI	5,000 PSI
6	None	28.27	56,548	70,685	84,822	113,090	141,370
	2-1/2	23.37	46,731	58,413	70,096	93,461	116,827
	3	21.21	42,411	53,014	63,616	84,822	106,027
	3-1/2	18.65	37,306	46,632	55,959	74,612	93,265
	4	15.71	31,416	39,270	47,124	62,832	78,540
7	None	38.49	76,970	96,213	115,455	153,940	192,425
	3	31.42	62,833	78,541	94,249	125,666	157,082
	3-1/2	28.87	57,728	72,160	86,592	115,456	144,320
	4	25.92	51,838	64,798	77,757	103,676	129,595
	4-1/2	22.58	45,162	56,453	67,743	90,324	112,905
	5	18.85	37,700	47,125	56,550	75,400	94,250
8	None	50.27	100,530	125,663	150,795	201,060	251,325
	3-1/2	40.64	81,288	101,610	121,932	162,576	203,220
	4	37.70	75,398	94,248	131,097	150,796	188,495
	4-1/2	34.36	68,722	85,903	103,083	137,444	171,805
	5	30.63	61,260	76,575	91,890	125,520	153,150
	5-1/2	26.51	53,014	66,268	79,521	106,028	132,535
10	None	78.54	157,080	196,350	235,620	314,160	392,700
	4-1/2	62.64	125,272	156,590	187,908	250,544	313,180
	5	58.91	117,810	147,263	176,715	235,620	294,525
	5-1/2	54.78	109,564	136,955	164,346	219,128	273,910
	7	40.06	80,110	100,138	120,165	160,220	200,275
12	None	113.1	226,200	282,750	339,300	452,400	565,500
	5-1/2	89.34	178,684	223,355	268,026	357,368	446,710
14	None	153.9	307,880	384,850	461,820	615,760	769,700
	7	115.5	230,910	288,638	346,365	461,820	577,275

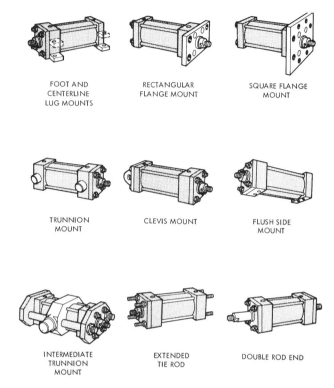

Figure 6–8 Cylinder mountings.

Trunnion Mount

In this configuration, trunnions or pivot rods are built into either the head or tail block of the cylinder. This type of mounting is used in applications that require movement of the cylinder body as well as that of the cylinder rod. Because both the cylinder body and rod move in this type of mounting configuration, great care must be taken to ensure that sideload and binding are prevented.

Clevis Mount

This type of mounting is similar to a rear-trunnion mount in that it permits the cylinder body to pivot at a point near the tail block. Unlike the trunnion, the clevis mounting arrangement cannot control the linear movement of the cylinder body. The looseness between the rear clevis and its mating mount and pin permits more uncontrolled side movement than in the trunnion mount.

RACK-AND-PINION, PISTON-TYPE ROTARY ACTUATOR

The rack-and-pinion-type actuators, also referred to as limited rotation cylinders, of the single or multiple bidirectional type are used for turning, positioning, steering,

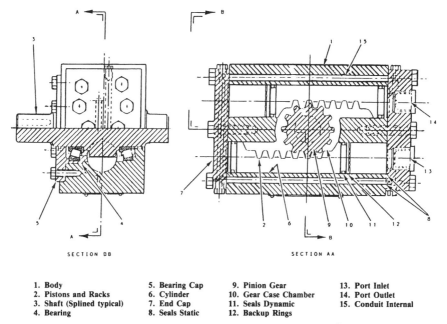

1. Body	5. Bearing Cap	9. Pinion Gear	13. Port Inlet
2. Pistons and Racks	6. Cylinder	10. Gear Case Chamber	14. Port Outlet
3. Shaft (Splined typical)	7. End Cap	11. Seals Dynamic	15. Conduit Internal
4. Bearing	8. Seals Static	12. Backup Rings	

Figure 6–9 Rack-and-pinion, double-piston rotary actuator.

opening, closing, swinging, or any other mechanical function involving restricted rotation.

Figure 6-9 shows a typical rack-and-pinion, double-piston actuator. The actuator consists of a body and two reciprocating pistons with an integral rack for rotating the shaft mounted in roller or journal bearings. The shaft and bearings are located in a central position and are enclosed with a bearing cap. The pistons, one on each side of the rack, are enclosed in cylinders machined or sleeved into the body. The body is enclosed with end caps and static seals to prevent external leakage of pressurized fluid.

In addition to its versatility, the cylinder-type actuator is probably the most trouble-free component of fluid power systems. However, it is very important that the cylinder, mechanical linkage, and actuating unit be correctly aligned. Any misalignment will cause excessive wear of the piston, piston rod, and seals. Also, proper adjustment between the piston rod and the actuating unit must be maintained.

HYDRAULIC MOTORS

A fluid power motor is a device that converts fluid power energy into rotary motion and force. The function of a motor is opposite that of a pump. However, the design and operation of fluid power motors are very similar to those of pumps. Therefore, a thorough knowledge of pumps will help you understand the operation of fluid power motors.

Motors have many uses in fluid power systems. In hydraulic power drives, pumps and motors are combined with suitable lines and valves to form hydraulic transmissions. The pump, commonly referred to as the A-end, is driven by some outside source such as an electric motor. The pump delivers pressurized fluid to the hydraulic motor, referred to as the B-end. The hydraulic motor is actuated by this flow and through mechanical linkage conveys rotary motion and force to do work.

Fluid motors may be either fixed or variable displacement. Fixed-displacement motors provide constant torque and variable speed. Controlling the amount of input flow varies the speed. Variable-displacement motors are constructed so that the working relationship of the internal parts can be varied to change displacement. The majority of the motors used in fluid power systems are the fixed-displacement type.

Although most fluid power motors are capable of providing rotary motion in either direction, some applications require rotation in only one direction. In these applications, one port of the motor is connected to the system pressure line and the other port to the return line. The flow of fluid to the motor is controlled by a flow control valve, by a two-way directional control valve, or by starting and stopping the power supply. Varying the rate of fluid flow to the motor may control the speed of the motor.

In most fluid power systems, the motor is required to provide actuating power in either direction. In these applications, the ports are referred to as working ports, alternating as inlet and outlet ports. Either a four-way directional control valve or a variable-displacement pump usually controls the flow to the motor.

Fluid motors are usually classified according to the type of internal element, which is directly actuated by the pressurized flow. The most common types of elements are gears, vanes, and pistons. All three of these types are adaptable for hydraulic systems, but only the vane type is used on pneumatic systems.

Gear-Type Motors

The spur, helical, and herringbone design gears are used in gear-type hydraulic motors. The motors use external-type gears, as discussed in the chapter on Pumps.

The operation of a gear-type motor is shown in Figure 6-10. Both gears are driven by gears. However, only one is connected to the output shaft. As pressurized fluid enters chamber A, it takes the path of least resistance and flows around the inside surface of the housing, forcing the gears to rotate as indicated. The flow continues through the outlet port to the return line. This rotary motion of the gears is transmitted through the attached shaft to the work unit.

The motor shown in Figure 6-10 is operating in one direction, but can be operated in either direction. To reverse the direction of rotation, the ports may be alternated as inlet and outlet. When fluid is directed through the outlet port into chamber B, the gears rotate in the opposite direction.

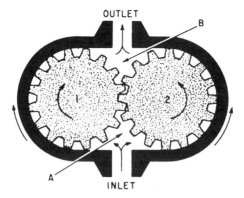

Figure 6–10 Gear-type hydraulic motor.

Vane-Type Motors

A typical vane-type motor is shown in Figure 6-11. This particular motor provides rotation in only one direction. The rotating element is a slotted rotor, which is mounted on a drive shaft. Each slot of the rotor is fitted with a free-sliding rectangular vane. The rotor and vanes are enclosed in the housing. The inner surface of the housing is offset from the drive shaft axis. When the rotor is in motion, the vanes tend to slide outward due to centrifugal force. The shape of the rotor housing limits the distance the vanes slide.

This motor operates on the principle of differential areas. When pressurized fluid is directed into the inlet port, its pressure is exerted equally in all directions. Since area A is greater than area B, the rotor will turn counterclockwise. Each vane in turn assumes the number 1 and number 2 positions, and the rotor turns continuously. The potential energy of the hydraulic fluid is thus converted into kinetic energy in the form

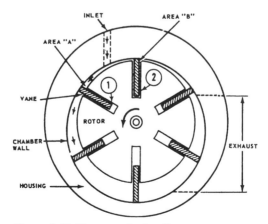

Figure 6–11 Vane-type hydraulic motor.

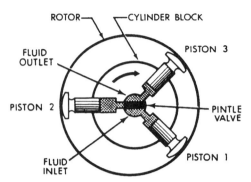

Figure 6–12 Typical piston motors.

of rotary motion and force. Many vane-type motors are capable of providing rotation in either direction. The two ports may be alternately used as inlet and outlet.

Piston-Type Motors

Piston-type motors are the most commonly used in hydraulic systems. They are basically the same as hydraulic pumps, except they are used to convert hydraulic energy into mechanical (rotary) energy. See Figure 6-12.

The most commonly used hydraulic motor is the fixed-displacement piston type. Some equipment uses a variable-displacement motor where very wide speed ranges are desired. Although directional control valves control some piston-type motors, they are often used in combination with variable-displacement pumps. This pump–motor combination is used to provide a transfer of power between a driving element and a driven element. Some applications for which hydraulic transmissions may be used are speed or constant-torque drives. The advantages of hydraulic transmission of power over mechanical transmission of power are as follows:

1. Quick, easy speed adjustments over a wide range, while the power source is operating at a constant, more efficient speed
2. Rapid, smooth acceleration or deceleration
3. Control over maximum torque and power
4. Smoother reversal of motion

7

CONTROL VALVES

It is impossible to design a practical fluid power system without some means of controlling the volume and pressure of the fluid, and directing that flow to the proper operating units. This is accomplished by the inclusion of control valves in the hydraulic circuit.

A valve is defined as any device by which the flow of fluid may be started, stopped, regulated, or directed by a movable part that opens or obstructs passage of the fluid. Valves must be able to accurately control fluid flow and system pressure, and to sequence the operation of all actuators within a hydraulic system.

Hydraulic control values can utilize a variety of actuators that activate their function. Normally these actuators use manual, electrical, mechanical, or pneumatic power sources.

VALVE CLASSIFICATIONS

Valves are classified by their intended use: flow control, pressure control, and direction control. Some valves have multiple functions that fall into more than one classification.

Flow Control Valves

Flow control valves are used to regulate the flow of fluids. Control of flow in hydraulic systems is critical because the rate of movement of fluid-powered machines or actuators depends on the rate of flow of the pressurized fluid. Some of the major types of flow control valves include:

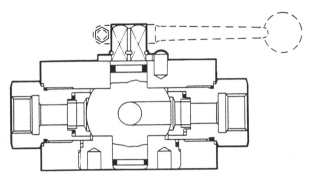

Figure 7–1   Typical ball valve.

Ball Valves

Ball valves are shutoff valves that use a ball to stop or start the flow of fluid down-stream of the valve. The ball, shown in Figure 7-1, performs the same function as the disc in other valves. As the valve handle is turned to open the valve, the ball rotates to a point where part or the entire hole that is machined through the ball is in line with the valve body inlet and outlet. This allows fluid flow to pass through the valve. When the ball is rotated so that the hole is perpendicular to the flow path, the flow stops.

Most ball valves are the quick-acting type. They require a 90-degree turn of the actua-tor lever to either fully open or completely close the valve. This feature, coupled with the turbulent flow generated when the ball opening is partially open, limits the use of ball valves as a flow control device. This type of valve is normally limited to strictly an "on–of" control function.

Gate Valves

Gate valves are used when a straight-line flow of fluid and minimum flow restriction are needed. Gate valves use a sliding plate within the valve body to stop, limit, or per-mit full flow of fluids through the valve. The gate is usually wedge-shaped. When the valve is wide open, the gate is fully drawn into the valve bonnet. This leaves the flow passage through the valve fully open with no flow restrictions. Therefore, there is lit-tle or no pressure drop or flow restriction through the valve.

Gate valves are not suitable for throttling volume. The control of flow is difficult because of the valve's design and the flow of fluid slapping against a partially open gate can cause extensive damage to the valve. Except as specifically authorized by the manufacturer, gate valves should not be used for throttling.

Gate valves are classified as either rising-stem or non-rising-stem valves. The non-rising-stem valve is shown in Figure 7-2. The stem is threaded into the gate. As the handwheel on the stem is rotated, the gate travels up or down the stem on the threads while the stem remains vertically stationary. This type of valve will almost always

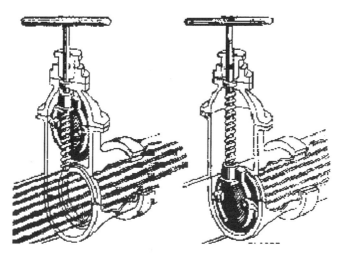

Figure 7–2 Operation of a gate valve.

have a pointer indicator threaded onto the upper end of the stem to indicate the position of the gate.

Valves with rising stems (Figure 7-3), are used when it is important to know by immediate inspection whether the valve is open or closed or when the threads exposed to the fluid could become damaged by fluid contamination. In this valve, the stem rises out of the valve bonnet when the valve is opened.

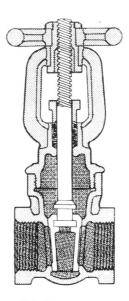

Figure 7–3 Rising stem gate valve.

Globe Valves

Globe valves are probably the most common valves in existence. The globe valve gets its name from the globular shape of the valve body. Other types of valves may also have globular bodies. Thus, it is the internal structure of the valve that defines the type of valve.

The inlet and outlet openings for globe valves are arranged in a way to satisfy the flow requirements. Figure 7-4 shows straight-, angle-, and cross-flow valves.

The part of the globe valve that controls flow is the disc, which is attached to the valve stem. Turning the valve stem in until the disc is seated into the valve seat closes the valve. This prevents fluid from flowing through the valve (Figure 7-5, view A). The edge of the disc and the seat are very accurately machined so that they form a tight seal when the valve is closed. When the valve is open (Figure 7-5, view B), the fluid flows through the space between the edge of the disc and the seat. Since the fluid flow is equal on all sides of the center of support when the valve is open, there is no unbalanced pressure on the disc that would cause uneven wear. The rate at which fluid flows through the valve is regulated by the position of the disc in relation to the valve seat. This type of valve is commonly used as a fully open or fully closed valve, but it may be used as a throttling valve. However, since the seating surface is a relatively large area, it is not suitable for a throttling valve where fine adjustment is required.

The globe valve should never be jammed in the open position. After a valve is fully opened, the handwheel or actuating handle should be turned toward the closed posi-

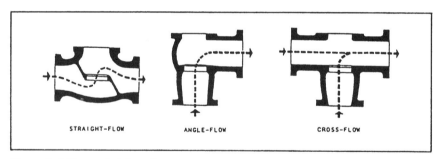

Figure 7–4 Types of globe valves.

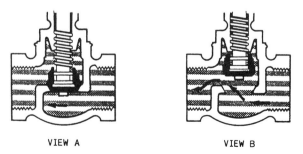

VIEW A   VIEW B

Figure 7–5 Operation of a globe valve.

tion approximately one-half turn. Unless this is done, the valve is likely to seize in the open position, making it difficult, if not impossible, to close the valve. Many valves are damaged in this manner. Another reason for not leaving globe valves in the fully open position is that it is sometimes difficult to determine if the valve is open or closed. If the valve is jammed in the open position, the stem may be damaged or broken by someone who thinks the valve is closed.

It is important that globe valves be installed with the pressure against the face of the disc to keep the system pressure away from the stem packing when the valve is shut.

Needle Valves

Needle valves are similar in design and operation to globe valves. Instead of a disc, a needle valve has a long tapered point at the end of the valve stem. Figure 7-6 shows a cross-sectional view of a needle valve.

The long taper of the valve element permits a much smaller seating surface area than that of the globe valve. Therefore, the needle valve is more suitable as a throttling valve. Needle valves are used to control flow into delicate gauges, which might be damaged by sudden surges of fluid flow under pressure.

Needle valves are also used to control the end of a work cycle, where it is desirable for motion to be brought slowly to a halt, and at other points where precise adjustments of flow rate are necessary and where a small rate of flow is desired.

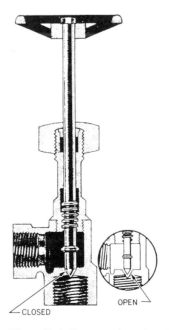

Figure 7–6 Cross-section of needle valve.

Although many of the needle valves used in fluid power systems are the manually operated types (Figure 7-6), modifications of this type of valve are often used as variable restrictors. This valve is constructed without a handwheel and is adjusted to provide a specific rate of flow. This rate of flow will provide a desired time of operation for a particular subsystem. Since this type of valve can be adjusted to conform to the requirements of a particular system, it can be used in a variety of systems. Figure 7-7 illustrates a needle valve that was modified as a variable restrictor.

Pressure Control

The safe and efficient operation of fluid power systems, system components, and related equipment requires a means to control pressure within the system. There are many types of automatic pressure control valves. Some of them merely provide an escape for excess pressures; some only reduce the pressure; and some keep the pressure within a preset range.

Relief Valves

Some fluid power systems, even when operated normally, may temporarily develop excessive pressure. For example, when an unusually strong work resistance is encountered, system pressure may exceed design limits. Relief valves are used to control this excess pressure.

Relief valves are automatic valves used on system lines and equipment to prevent overpressurization. Most relief valves simply open at a preset pressure and shut when the pressure returns to normal. They do not maintain flow or pressure at a given level; they simply prevent pressure from rising above a specified limit.

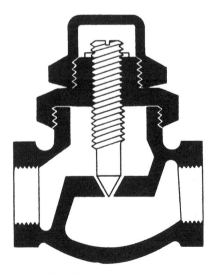

Figure 7–7 Variable restrictor.

Main system relief valves are generally installed between the pump or pressure source and the first system isolation valve. The valve must be large enough to allow the full output of the hydraulic pump to be delivered back to the reservoir. This design feature, called a full-flow bypass, is essential for all hydraulic systems. The location of the valve is also critical. If the valve were installed downstream from the system isolator valve, the pump could be deadheaded when the system was shut down.

Smaller relief valves are often used in isolated parts of the system where a check valve or directional control valve prevents pressure from being relieved through the main system relief valve or where pressures must be relieved at a specific set point lower than the main system pressure. These small relief valves are also used to relieve pressures caused by thermal expansion of fluids.

Figure 7-8 shows a typical relief valve. System pressure simply acts under the valve disc at the inlet of the valve. When the system pressure exceeds the preload force exerted by the valve spring, the valve disc will lift off of its seat. This will allow some of the system fluid to escape through the valve outlet. Flow will continue until the system pressure is reduced to a level below the spring force.

All relief valves have an adjustment for increasing or decreasing the set relief pressure. Some relief valves are equipped with an adjusting screw for this purpose. This adjusting screw is usually covered with a cap, which must be removed before an adjustment can be made.

Some type of locking device, such as a lock nut, is usually provided to prevent the adjustment from changing through vibration. Other types of relief valves are equipped

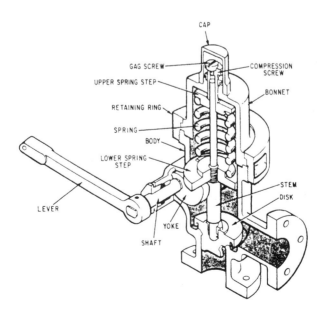

Figure 7–8 Cutaway of relief valve.

with a handwheel for making adjustments of the valve. Either the adjusting screw or the handwheel is turned clockwise to increase the pressure at which the valve will open. In addition, most relief valves are also provided with an operating lever or some type of device to allow manual cycling or opening of the valve for certain tasks, such as testing and inspection.

Various modifications of the relief valve shown in Figure 7-8 are used to efficiently serve the requirements of some fluid power applications. This relief valve is unsatisfactory for some applications. To provide a better understanding of the operation of relief valves, we will discuss some of the undesirable characteristics of this valve.

A simple relief valve, such as the one illustrated in Figure 7-8, with a suitable spring adjustment can be set so that it will open when the system pressure reaches a certain level, 500 psi, for example. When the valve opens, the volume of flow to be handled may be greater than the capacity of the valve. If this happens, the system pressure may be increased by several hundred psi above the set pressure.

A simple relief valve will only be effective when it is sized to permit full flow of fluid at its point of installation. In undersized valves, the valve element will chatter back and forth and create turbulent flow through the valve.

The surface area of the valve element must be larger than that of the pressure opening. In the case of the valve illustrated in Figure 7-9, the force exerted directly upward by system pressure when the valve is closed depends on the area A across the valve element where the element seats against the pressure tube. The moment the valve opens, the upward force exerted depends on the horizontal area B of the entire valve element, which is greater than area A. This causes an immediate upward jump of the valve element as the element leaves the valve seat. It also requires a greater force to close the valve than was required to open it. As a result, the valve will not close until the system pressure has decreased to a certain point below the pressure required to open it.

Let us assume that a relief valve of this type is set to open at 500 psi. When the valve is closed, the pressure acts on area A. If this area A is 0.5 square inch, an upward force of 250 pounds (500 × 0.5) will be exerted on the valve at the moment of opening.

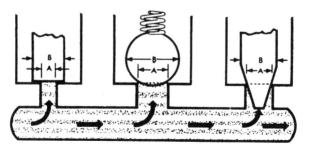

Figure 7–9 Pressure acting on different areas.

With the valve open, the pressure acts on area B. If area B is 1 square inch, the upward force is 500 psi, or double the force at which the valve came off its seat. To close the valve system, pressure would have to decrease well below the point at which the valve opened, or less than 250 psi. The exact pressure will depend on the shape of the valve element.

In some hydraulic systems, there is a pressure in the return line. This back pressure is caused by restrictions in the return line and will vary in relation to the amount of fluid flowing in the return line. This pressure creates a force on the back of the valve element and will increase the force necessary to open the valve and relieve system pressure.

It follows that simple relief valves have a tendency to open and close rapidly as they "hunt" above and below the set pressure. This causes pressure pulsations and undesirable vibration. Because of the unsatisfactory performance of the simple relief valve in some applications, compound relief valves were developed.

Compound relief valves use the principles of operation of simple relief valves for one stage of their action—that of a pilot valve. Provision is made to limit the amount of fluid that the pilot valve must handle, which avoids the weakness of the simple valve.

The operation of a compound relief valve is illustrated in Figure 7-10. In view A, the main valve, which consists of a piston, stem, and spring, is closed, blocking flow from the high-pressure line to the reservoir. Fluid in the high-pressure line flows around the stem of the main valve as it flows to the actuating unit. The stem of the main valve is hollow and contains the main valve spring, which forces the main valve against its seat. When the pilot valve is open, the stem passage allows fluid to flow from the pilot valve, around the main valve spring, and down to the return line.

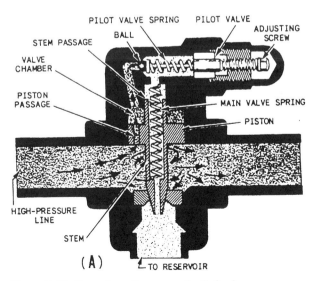

Figure 7–10 Operation of compound relief valve.

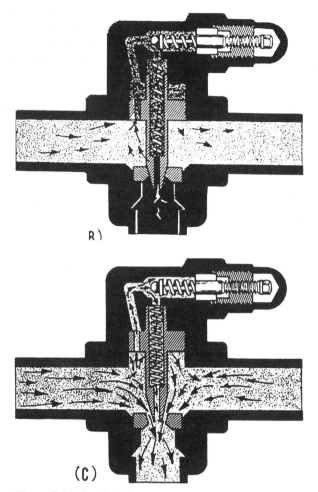

Figure 7–10 Continued.

There is also a narrow passage (piston passage) through the main valve piston. This passage connects the high-pressure line to the valve chamber.

The pilot valve is a small, ball-type, spring-loaded check valve. It connects the top of the passage from the valve chamber with the passage through the main valve stem. The pilot valve is the control unit of the relief valve because the pressure at which the relief valve will open depends on the tension of the pilot valve spring. The pilot valve spring tension is adjusted by turning the adjusting screw so that the ball will unseat when system pressure reaches the preset limit.

Fluid at line pressure flows through the narrow piston passage to fill the chamber. Because the line and the chamber are connected, the pressures in both are equal. The top and bottom of the main piston have equal areas. Therefore, the hydraulic forces acting upward and downward are equal and there is no tendency for the piston to move in either direction. The only other force acting on the main valve is that of the main valve spring, which holds it closed.

When the pressure in the high-pressure line increases to the point at which the pilot valve is set, the ball unseats (Figure 7-10, view B). This opens the valve chamber through the valve stem passage to the low-pressure return line. Fluid immediately begins to flow out of the chamber much faster than it can flow through the narrow piston passage. As a result, the chamber pressure immediately drops, and the pilot valve begins to close again, restricting the outward flow of fluid. Chamber pressure therefore increases, the valve opens, and the cycle repeats.

So far, the only part of the valve that has moved appreciably is the pilot, which functions just like any other simple spring-loaded relief valve. Because of the small size of the piston passage, there is a severe limit on the amount of overpressure protection the pilot can provide to the system. All the pilot valve can do is limit fluid pressure in the valve chamber above the main piston to a preset maximum pressure, by allowing excess fluid to flow through the piston passage, through the stem passage, and into the return line. When pressure in the system increases to a value that is above the flow capacity of the pilot valve, the main valve opens, permitting excess fluid to flow directly to the return line. This is accomplished in the following manner.

As system pressure increases, the upward force on the main piston overcomes the downward force. The downward force consists of the main piston spring tension and the fluid pressure in the valve chamber (Figure 7-10, view C). The piston then rises, unseating the stem and allowing the fluid to flow from the system pressure line directly into the return line. This causes the system pressure to decrease rapidly, since the main valve is designed to handle the complete output of the pump. When the pressure returns to normal, the pilot spring forces the ball onto its seat. Pressures are equal above and below the main piston, and the main spring forces the valve to seat.

As you can see, the compound valve overcomes the greatest limitation of a simple relief valve by limiting the flow through the pilot valve to the quantity it can satisfactorily handle. This limits the pressure above the main valve and enables the main-line pressure to open the main valve. In this way, the system is relieved when an overload exists.

Pressure Regulators

Pressure regulators, often referred to as unloading valves, are used in fluid power systems to regulate pressure. In hydraulic systems the pressure regulator is used to unload the pump and to maintain, or regulate, system pressure at the desired values.

Not all hydraulic systems require pressure regulators. The open-center system does not require such a regulator. Many systems are equipped with variable-displacement pumps, which contain a pressure-regulating device.

Pressure regulators are made in a variety of types. However, the basic operating principles of all regulators are similar to those of the one illustrated in Figure 7-11.

A regulator is open when it is directing fluid under pressure into the system (Figure 7-11, view A). In the closed position (Figure 7-11, view B), the fluid in the part of the system beyond the regulator is trapped at the desired pressure and the fluid from the pump is bypassed into the return line and back to the reservoir. To prevent constant opening and closing (chatter), the regulator is designed to open at pressure somewhat lower than the closing pressure. This difference is known as differential or operating range. For example, assume that a pressure regulator is set to open when the system pressure drops below 600 psi and close when the pressure rises above 800 psi. The differential or operating range is 200 psi.

Referring to Figure 7-11, assume that the piston has an area of 1 square inch, the pilot valve has a cross-sectional area of one-fourth (1/4) square inch, and the piston spring provides 600 pounds of force that pushes the piston against its seat. When the system pressure is less than 600 psi, fluid from the pump will enter the inlet port and flow to the top of the regulator, then to the pilot valve. When the system pressure at the valve inlet increases to the point where the force it creates against the front of the check valve exceeds the force created against the back of the check valve, the check valve opens. This allows fluid to flow into the system and to the bottom of the regulator against the piston. When the system force exceeds the force exerted by the spring, the piston moves up, causing the pilot valve to unseat. Since the fluid will take the path of least resistance, it will pass through the regulator and back to the reservoir through the bypass line.

When the fluid from the pump is suddenly allowed a free path to return, the pressure on the input side of the check valve drops and the check valve closes. The fluid in the system is then pressurized until a power unit is actuated or until pressure is slowly lost through normal internal leakage within the system.

When the system pressure decreases to a point slightly below 600 psi, the spring forces the piston down and closes the pilot valve. When the pilot valve is closed, the fluid cannot flow directly to the return line. This causes the pressure to increase in the

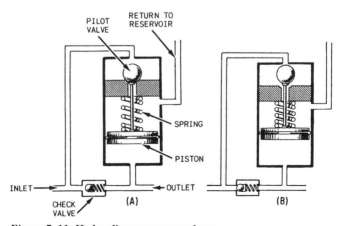

Figure 7–11 Hydraulic pressure regulator.

line between the pump and the regulator. This pressure opens the check valve, causing fluid to enter the system.

In summary, when the system pressure decreases, the pressure regulator will open, sending fluid to the system. When the system pressure increases, the regulator will close, allowing the fluid from the pump to flow through the regulator and back to the reservoir. The pressure regulator takes the load off of the pump and regulates system pressure.

SEQUENCE VALVES

Sequence valves control the sequence of operation between two branches in a hydraulic circuit. In other words, they enable one component within the system to automatically set another component into motion. An example of the use of a sequence valve is in an aircraft landing gear actuating system.

In a landing gear actuating system, the landing gear doors must open before the landing gear starts to extend. Conversely, the landing gear must be completely retracted before the doors close. A sequence valve installed in each landing gear actuating line performs this function.

A sequence valve is somewhat similar to a relief valve except that, after the set pressure has been reached, the sequence valve diverts the fluid to a second actuator or motor to do work in another part of the system. Figure 7-12 shows an installation of two sequence valves that control the sequence of operation of three actuating cylinders. Fluid is free to flow into cylinder A. The first sequence valve (1) blocks the passage of fluid until the piston in cylinder A moves to the end of its stroke. At this time, sequence valve 1 opens, allowing fluid to enter cylinder B. This action continues until all three pistons complete their strokes.

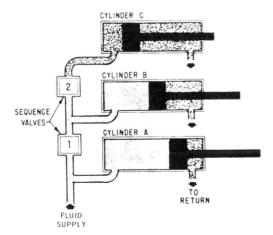

Figure 7–12 Installation of sequence valve.

There are various types of sequence valves. Some are controlled by pressure and some are controlled mechanically.

Pressure-Controlled Sequence Valves

The operation of a typical pressure-controlled sequence valve is illustrated in Figure 7-13. The opening pressure is obtained by adjusting the tension of the spring that normally holds the piston in the closed position. Fluid enters the valve through the inlet port, flows around the lower part of the piston, then exits through the outlet port and flows to the primary unit. This fluid pressure also acts against the lower surface of the piston. Since the top of the piston has a larger cross-sectional area than the bottom, this flow does not cause the piston to lift.

When the primary actuating unit completes its operation, pressure in the line to the actuating unit increases sufficiently to overcome the force of the sequence valve spring and the piston rises. The valve is then in the open position. The fluid entering the valve takes the path of least resistance and flows to the secondary units.

A drain passage is provided to allow any fluid leaking past the piston to flow from the top of the valve. In hydraulic systems, this drain line is usually connected to the main return line.

Mechanically Operated Sequence Valves

The mechanically operated sequence valve (Figure 7-14) is operated by a plunger that extends through the valve body. The valve is mounted so that the primary unit will operate the plunger.

A check valve, either a ball or poppet, is installed between the fluid ports in the body. Either the plunger or fluid pressure can unseat it. Port A and the actuator of the primary unit are connected to a common line. Port B is connected by a line to the actuator of the secondary unit. When fluid under pressure flows to the primary unit, it also flows into the sequence valve. In order to operate the secondary unit, the fluid must flow through

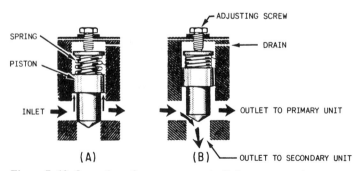

Figure 7–13 Operation of a pressure-controlled sequence valve.

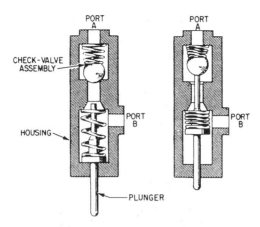

Figure 7–14 Mechanically operated sequence valve.

the sequence valve. The valve is located so that the primary unit depresses the plunger as it completes its operation. The plunger unseats the check valve and allows the fluid to flow through the valve, out port B, and to the secondary unit.

This type of sequence valve permits flow in the opposite direction. Fluid enters port B and flows to the check valve. Although this is return flow from the actuating unit, the fluid overcomes spring tension, unseats the check valve, and flows through port A.

PRESSURE-REDUCING VALVES

Pressure-reducing valves provide a steady pressure into a part of the system that operates at a pressure lower that normal system pressure. A reducing valve can normally be set for any desired downstream pressure within its design limits. Once the valve is set, the reduced pressure will be maintained regardless of changes in the supply pressure and system load variations.

There are various designs and types of pressure-regulating valves. The spring-loaded reducer and the pilot-controlled valve are the most common.

Spring-Loaded Pressure-Reducing Valves

The spring-loaded pressure-reducing valve (Figure 7-15) is commonly used in pneumatic systems. It is often referred to as a pressure regulator. The valve simply uses spring pressure against a diaphragm to open the valve. On the bottom of the diaphragm, the outlet pressure of the valve forces the diaphragm upward to shut the valve. When the outlet pressure drops below the set point of the valve, the spring pressure overcomes the outlet pressure and forces the valve stem downward, opening the valve. As the outlet pressure increases, approaching the desired pressure, the pressure under the diaphragm begins to overcome spring pressure, forcing the valve stem upward and closing the valve. You can adjust the downstream pressure by turning the

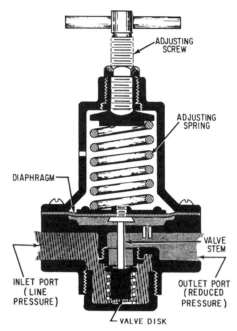

Figure 7–15 Spring-loaded pressure-reducing valve.

adjusting screw, which varies the spring pressure against the diaphragm. This particular spring-loaded valve will fail in the open position if a diaphragm rupture occurs.

Pilot-Controlled Pressure-Reducing Valves

Figures 7-16A and B illustrate the operation of a pilot-controlled valve. This valve consists of an adjustable pilot valve that controls the operating pressure of the valve, and a spool valve that reacts to the action of the pilot valve.

The pilot valve consists of a poppet (1), a spring (2), and an adjusting screw (3). The valve spool assembly consists of: a valve spool (10) and a spring (4). Fluid under main system pressure enters the inlet port (11) and under all conditions is free to flow through the valve and the outlet port (5).

Figure 7-16, view A, shows the valve in the open position. In this position, the pressure in the reduced-pressure outlet port (6) has not reached the preset operating pressure of the valve. The fluid also flows through passage 8, through smaller passage 9 in the center of the valve spool, and into chamber 12. The fluid pressure at outlet port 6 is therefore distributed to both ends of the spool. When these pressures are equal, the spool is hydraulically balanced. Spring 4 is a low-tension spring and applies only a slight downward force on the spool. Its main purpose is to position the spool and to maintain opening 7 at its maximum size.

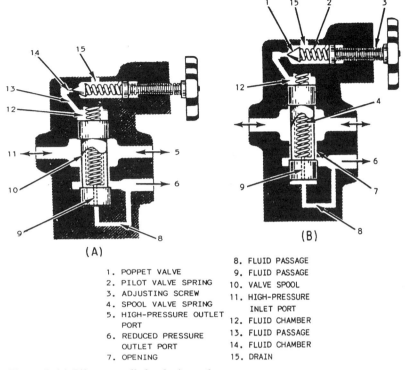

1. POPPET VALVE
2. PILOT VALVE SPRING
3. ADJUSTING SCREW
4. SPOOL VALVE SPRING
5. HIGH-PRESSURE OUTLET PORT
6. REDUCED PRESSURE OUTLET PORT
7. OPENING
8. FLUID PASSAGE
9. FLUID PASSAGE
10. VALVE SPOOL
11. HIGH-PRESSURE INLET PORT
12. FLUID CHAMBER
13. FLUID PASSAGE
14. FLUID CHAMBER
15. DRAIN

Figure 7–16 Pilot-controlled reducing valve.

As the pressure increases in outlet port 6, this pressure is transmitted through passages 8 and 9 to chamber 12. This pressure also acts on the pilot valve poppet (1). When this pressure increases above the preset operating pressure, it overcomes the force of the pilot valve spring 2 and unseats the poppet. This allows fluid to flow through the drain port (15). Because the small passage 9 restricts flow into chamber 12, the fluid pressure in the chamber drops. This causes a momentary differential in pressure across the valve spool 10, which allows fluid pressure acting against the bottom area of the valve spool to overcome the downward force of spring 4. The spool is then forced upward until the pressure across its ends is equalized. As the spool moves upward, it restricts the flow through opening 7 and causes the pressure to decrease in the reduced pressure outlet, port 6. If the pressure in the outlet port continues to increase to a value above the preset pressure, the pilot valve will open again and the cycle will repeat. This allows the spool valve to move up higher into chamber 12, thus further reducing the size of opening 7. These cycles repeat until the desired pressure is maintained in outlet 6.

When the pressure in outlet 6 decreases to a value below the preset pressure, spring 4 forces the spool downward, allowing more fluid to flow through opening 7.

COUNTERBALANCED VALVES

The counterbalance valve is normally located in the line between a directional control valve and the outlet of a vertical mounted actuating cylinder that supports weight or must be held in position for a period of time. This valve serves as a hydraulic resistance to the actuating cylinder. For example, counterbalance valves are used in some hydraulically operated forklifts. The valve offers a resistance to the flow from the actuating cylinder when the fork is lowered. It also helps to support the fork in the UP position.

One type of counterbalance valve is illustrated in Figure 7-17. The valve element is a balanced spool (4). The spool consists of two pistons permanently fixed on either end of a shaft. The inner surface areas of the pistons are equal. Therefore, pressure acts equally on both areas regardless of the position of the valve and has no effect on the movement of the valve. The shaft area between the two pistons provides the area for the fluid to flow when the valve is open. A small piston (9) is attached to the bottom of the spool valve.

When the valve is in the closed position, the top piston of the spool blocks the discharge port (8). With the valve in this position, fluid flowing from the actuating unit enters the inlet port (5). The fluid cannot flow through the valve because the discharge port (8) is blocked. However, fluid will flow through the pilot passage (6) to the small pilot piston. As the pressure increases, it acts on the pilot piston until it overcomes the preset pressure of spring 3.

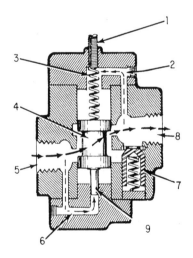

1. ADJUSTMENT SCREW	6. PILOT PASSAGE
2. INTERNAL DRAIN	7. CHECK VALVE
3. SPRING	8. DISCHARGE OUTLET
4. SPOOL	OR REVERSE FREE
5. PRESSURE INLET OR	FLOW INLET
REVERSE FREE FLOW	9. PISTON
OUTLET	

Figure 7–17 Counterbalance valve.

This forces the valve spool (4) up and allows the fluid to flow around the shaft of the valve spool and out discharge port 8. Figure 7-17 shows the valve in this position. During reverse flow, the fluid enters port 8. The spring (3) forces valve spool 4 to the closed position. The fluid pressure overcomes the spring tension of check valve 7. The check valve opens and allows free flow around the shaft of the valve spool and out through port 5.

The operating pressure of the valve can be adjusted by turning the adjustment screw (1), which increases or decreases the tension of the spring. This adjustment depends on the weight that the valve must support. It is normal for a small amount of fluid to leak around the top piston of the spool valve and into the area around the spring. An accumulation would cause additional pressure on top of the spool valve. This would require additional pressure to open the valve. The drain (2) provides a passage for this fluid to flow to port 8.

DIRECTIONAL CONTROL VALVES

Directional control valves are designed to direct the flow of fluid, at the desired time, to the point in a fluid power system where it will do work. The driving of a ram back and forth in its cylinder is an example of when a directional control valve is used. Various other terms are used to identify directional valves, such as selector valve, transfer valve, and control valve. This manual will use the term directional control valve to identify these valves.

Directional control valves for hydraulic and pneumatic systems are similar in design and operation. However, there is one major difference. The return port of a hydraulic valve is ported through a return line to the reservoir. Any other differences are pointed out in the discussion of these valves.

Directional control valves may be operated by differences in pressure acting on opposite sides of the valve elements, or they may be positioned manually, mechanically, or electrically. Often two or more methods of operating the same valve will be used in different phases of its action.

Directional control valves can be classified in several ways: by the type of control, the number of ports in the valve housing, and the specific function that the valve performs. The most common method is by the type of valving element used in the construction of the valve. The most common types of valving elements are the ball, cone, sleeve, poppet, rotary spool, and sliding spool. The basic operating principles of the poppet, rotary spool, and sliding spool types are discussed in this text.

The poppet fits into the center bore of the seat (Figure 7-18). The seating surfaces of the poppet and the seat are lapped or closely machined so that the center bore will be sealed when the poppet is seated (shut). The action of the poppet is similar to that of the valves in an automobile engine. In most valves the poppet is held in the seated position by a spring.

The valve consists primarily of a movable poppet, which closes against the valve seat. In the closed position, fluid pressure on the inlet side tends to hold the valve tightly

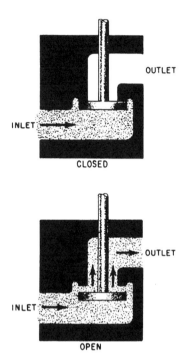

Figure 7–18 Operation of a simple poppet valve.

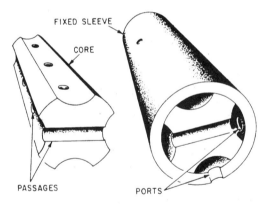

Figure 7–19 Parts of a rotary spool valve.

closed. A small amount of movement from a force applied to the top of the poppet stem opens the poppet and allows fluid to flow through the valve.

The rotary spool directional control valve (Figure 7-19) has a round core with one or more passages or recesses in it. The core is mounted within a stationary sleeve. As the core is rotated within the stationary sleeve, the passages or recesses connect or block the ports in the sleeve. The ports in the sleeve are connected to the appropriate lines of the fluid system.

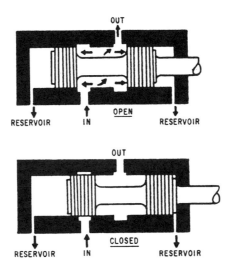

Figure 7–20 Two-way sliding spool valve.

The operation of a simple sliding spool directional control valve is shown in Figure 7-20. The valve is so named because of the shape of the valving element that slides back and forth to block or open ports in the valve housing. The sliding element is referred to as the spool or piston. The inner piston areas, or lands, are equal. Thus, fluid under pressure, which enters the valve from the inlet, ports acts equally on both inner piston areas, regardless of the position of the spool. Sealing is usually accomplished by a very closely machined fit between the spool and the valve body or sleeve. For valves with more ports, the spool is designed with more pistons or lands on a common shaft. The sliding spool is the most common type of directional control valve.

CHECK VALVES

Check valves are used in fluid systems to permit flow in one direction and to prevent flow in the opposite direction. They are classified as one-way directional control valves. The check valve may be installed independently in a line to allow one-direction flow, or it may be used as an integral part of globe, sequence, counterbalance, and pressure-reducing valves.

Check valves are available in various designs. They are opened by the force of fluid in motion flowing in one direction, and are closed by fluid attempting to flow in the opposite direction. The force of gravity or the action of a spring aids in closing the valve.

Figure 7-21 shows a swing check valve. In the open position, the flow of fluid forces the hinged disc up and allows free flow through the valve. Flow in the opposite direction, or loss of flow, with the aid of gravity forces the hinged disc to close the passage

Figure 7–21 Swing check valve.

and blocks the flow. This type of valve is sometimes designed with a spring to assist in closing the valve.

The most common type of check valve uses either a ball or cone for the sealing element (Figure 7-22). As fluid pressure is applied in the direction of the arrow (flow), the cone (view A) or ball (view B) is forced off its seat, allowing fluid to flow freely through the valve. This valve is known as a spring-loaded check valve.

The spring is installed in the valve to hold the cone or ball on its seat whenever fluid is not flowing. The spring also helps to force the cone or ball on its seat when the fluid attempts to flow in the opposite direction. Since the opening and closing of this type of valve are not dependent on gravity, its location in a system is not limited to the vertical position.

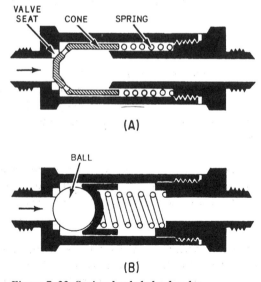

Figure 7–22 Spring-loaded check valve.

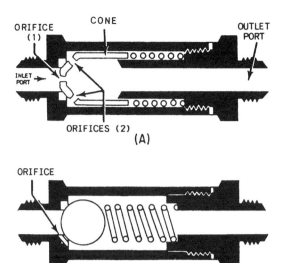

Figure 7–23 Typical orifice check valve.

A modification of the spring-loaded check valve is the orifice check valve (Figure 7-23A,B). This valve allows normal flow in one direction and restricted flow in the opposite direction. It is referred to as a one-way restrictor.

Figure 7-23, view A, shows a cone-type orifice check valve. When sufficient fluid pressure is applied at the inlet port, it overcomes spring tension and moves the cone off of its seat. The two orifices (2) in the illustration represent several openings located around the slanted circumference of the cone. These orifices allow free flow of fluid through the valve while the cone is off its seat. When fluid pressure is applied through the outlet port, the force of the fluid and spring tension move the cone to the left and onto its seat. This action blocks the flow of fluid though the valve, except through the orifice (1) in the center of the cone. The size of the orifice determines the rate of flow through the valve as the fluid flows from right to left.

SHUTTLE VALVES

In certain fluid power systems, the supply of fluid to subsystems must be from more than one source to meet system requirements. In some systems an emergency system is provided as a source of pressure in the event of normal system failure. The emergency system will usually actuate only essential components.

The main purpose of the shuttle valve is to isolate the normal system from an alternate or emergency system. It is small and simple, yet it is a very important component.

Figure 7-24 is a cutaway view of a typical shuttle valve. The housing contains three ports: the normal system inlet, the secondary or emergency inlet, and the normal out-

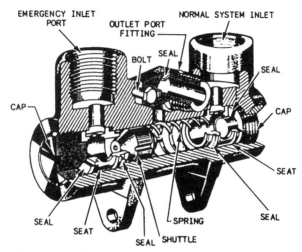

Figure 7–24 Shuttle valve.

let. A shuttle valve used to operate more than one actuating unit may contain additional unit outlet ports. Enclosed in the housing is a sliding part called the shuttle. Its purpose is to seal off one or the other of the inlet ports. There is a shuttle seat at each inlet port. When a shuttle valve is in the normal operating position, fluid has a free flow from the normal system inlet port, through the valve, and out through the outlet port.

In this position, the shuttle is seated against the alternate system inlet port and held there by system pressure and the shuttle valve spring. The shuttle remains in this position until the alternate system is activated. This action directs fluid under pressure from the alternate source to the shuttle valve and forces the shuttle from the alternate inlet port seat to the normal inlet port seat. Fluid from the alternate system then has a free flow path to the outlet port and prevents inlet flow from the primary or normal system.

The shuttle may be one of four types: (1) sliding plunger, (2) spring-loaded piston, (3) spring-loaded ball, or (4) spring-loaded poppet. In shuttle valves that are designed with a spring, the shuttle is normally held against the alternate system inlet port by the spring.

TWO-WAY VALVES

The term two-way indicates that the valve contains and controls two functional flow-control ports. A two-way, sliding spool directional control valve is shown in Figure 7-25. As the spool is moved back and forth, it either allows fluid to flow through the valve or prevents flow. In the open position, the fluid enters the inlet port and flows around the shaft of the spool, then through the outlet port. The spool cannot move back and forth by difference of forces set up within the cylinder, since the forces there

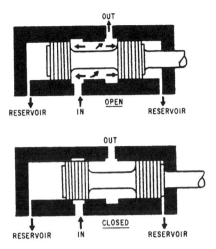

Figure 7–25 Two-way sliding spool valve.

are equal. As indicated by the arrows against the pistons of the spool, the same pressure acts on equal areas on their inside surfaces. In the closed position, one of the pistons of the spool simply blocks the inlet port, thus preventing flow through the valve.

A number of features common to most sliding spool valves are shown in Figure 7-25. The small ports at either end of the valve housing provide a path for any fluid that leaks past the spool to flow to the reservoir. This prevents pressure from building up against the ends of the pistons, which would hinder the movement of the spool. When spool valves become worn, they may lose balance because of greater leakage on one side of the spool than on the other. In that event, the spool would tend to stick when it is moved back and forth. Small grooves are therefore machined around the sliding surface of the piston. In hydraulic valves, leaking liquid will encircle the piston and keep the contacting surfaces lubricated and centered.

THREE-WAY VALVES

Three-way valves contain a pressure port, a cylinder port, and a return or exhaust port. The three-way directional control valve is designed to operate an actuating unit in one direction. It permits either the load on the actuating unit or a spring to return the unit to its original position.

Cam-Operated Three-Way Valves

Figure 7-26A, B shows the operation of a cam-operated, three-way, poppet-type directional control valve.

View A shows fluid under pressure forcing the piston outward against a load. The upper poppet (2) is unseated by the inside cam (5), permitting fluid to flow from the

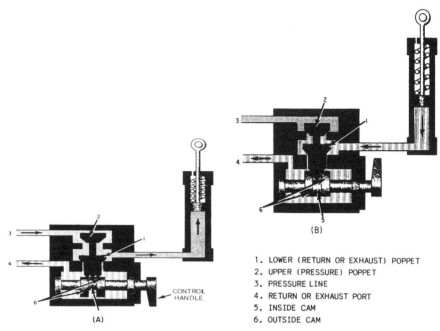

1. LOWER (RETURN OR EXHAUST) POPPET
2. UPPER (PRESSURE) POPPET
3. PRESSURE LINE
4. RETURN OR EXHAUST PORT
5. INSIDE CAM
6. OUTSIDE CAM

Figure 7–26 Three-way poppet-type control valve.

line (3) into the cylinder to actuate the piston. The lower poppet (1) is seated, sealing off the flow into the return line (4). As the force of the pressurized fluid extends the piston rod, it also compresses the spring in the cylinder.

View B shows the valve with the control handle turned to the opposite position. In this position, the upper poppet (2) is seated, blocking the flow of fluid from the pressure line (3). The lower poppet (1) is unseated by the outside cam (6). This releases the pressure in the cylinder and allows the spring to expand, which forces the piston rod to retract. The fluid from the cylinder flows through the control valve and out the return port (4). In hydraulic systems, a line to the reservoir connects the return port.

Pilot-Operated Three-Way Valves

A pilot-operated, poppet-type, three-way directional control valve is shown in Figure 7-27. This valve is normally closed and is forced open by fluid pressure entering the pilot chamber. The valve contains two poppets connected to each other by a common stem. The poppets are connected to diaphragms, which hold them in a centered position.

The pressure in the pilot port and the chamber above the upper diaphragm controls the movement of the poppet. When the pilot chamber is not pressurized, the lower poppet is seated against the lower valve seat. Fluid can flow from the supply line through the inlet port and through holes in the lower diaphragm to fill the bottom chamber. This

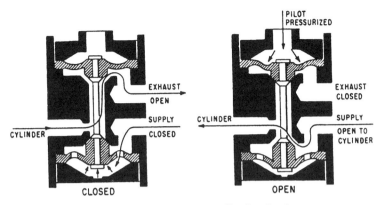

Figure 7–27   Three-way, poppet-type, normally closed valve.

pressure holds the lower poppet tightly against its seat and blocks flow from the inlet port through the valve. At the same time, because of the common stem, the upper poppet is forced off of its seat. Fluid from the actuating unit flows through the open passage, around the stem, and through the exhaust port.

When the pilot chamber is pressurized, the force acting against the diaphragm forces the poppet down. The upper poppet closes against its seat, blocking flow of fluid from the cylinder to the exhaust port. The lower poppet opens, and the passage from the supply inlet port to the cylinder port is open so that the fluid can flow to the actuating unit.

FOUR-WAY VALVES

Most actuating devices require system pressure for operation in two directions. The four-way directional control valve, which contains four ports, is used to control the operation of such devices. The four-way valve is also used in some systems to control the operation of other valves. It is one of the most widely used directional control valves in fluid power systems.

The typical four-way directional control valve has four ports: a pressure port, a return port, and two cylinder or work ports (output). The pressure port is connected to the main system pressure line and the return port to the return line to the reservoir. The two outputs are connected to the actuating unit.

Poppet-Type Four-Way Valves

Figure 7-28 shows a typical four-way valve, a poppet-type directional control valve. This is a manually operated valve and consists of a group of conventional spring-loaded poppets. The poppets are enclosed in a common housing and are interconnected by ducts to direct flow of fluid in the desired direction.

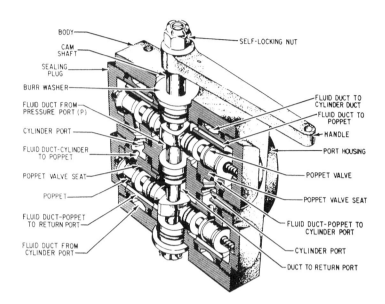

Figure 7–28 Cutaway view of poppet-type four-way valve.

Cams on a camshaft actuate the poppets. The movement of the manual handle controls the camshaft. The valve may be operated by manually moving the handle or mechanical linkage to a control handle, which is located in a convenient place for the operator.

The camshaft may be rotated to any one of three positions: neutral and two working positions. In the neutral position, the camshaft lobes are not contacting any of the poppets. This ensures that the poppet springs will hold all four poppets firmly seated. With all poppets seated, there is no flow through the valve. This also blocks the two cylinder or work ports, so that when the valve is in neutral, the fluid in the actuating unit is trapped. Relief valves are installed in both working lines to prevent overpressurization caused by thermal growth and in the main system pressure line to prevent damage to system components.

The poppets are arranged so that rotation of the camshaft will open the proper combination of poppets to direct flow through the desired working lines to an actuating unit. At the same time, fluid will be directed from the actuating unit through the opposite working line, through the valve, and back to the reservoir.

To stop rotation of the camshaft at an exact position, a stop pin is secured to the body and extends through a cutout section of the camshaft flange. This stop pin prevents over-travel by ensuring that the camshaft stops rotation at the point where the cam lobes have moved the poppet the greatest distance from their seats.

O-rings are spaced at intervals along the length of the shaft to prevent external leakage around the end of the shaft and internal leakage from one chamber to another. The

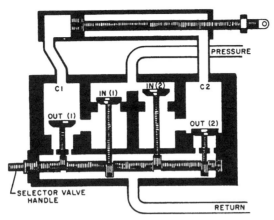

Figure 7–29 Working view of a poppet-type four-way valve.

camshaft has two lobes, or raised portions. The shape of these lobes is such that when the shaft is placed in the neutral position the lobes will not contact any of the poppets.

When the handle is moved in either direction from neutral, the camshaft is rotated. This rotates the lobes, which unseats one pressure poppet and one return poppet (Figure 7-29). The valve is now in the working position. Fluid under pressure, entering the pressure port, flows through the vertical fluid passage in both pressure poppets' seats.

Since the cam lobe unseats only one pressure poppet, IN (2), the fluid flows past the open poppet to the inside of the poppet seat. From there it flows through the diagonal passage, out one cylinder port, C2, and to the actuating unit.

Return fluid from the actuating unit enters the other cylinder port, C1. It then flows through the corresponding fluid passage, past the unseated return poppet, OUT (1), through the vertical fluid passage, and out the return port.

When the camshaft is rotated in the opposite direction to the neutral position, the two poppets seat and the flow stops. When the camshaft is rotated further in this direction until it hits the opposite stop pin, the opposite pressure and return poppets are unseated. This reverses the flow in the working lines, causing the actuating unit to move in the opposite direction.

Rotary Spool Four-Way Valves

Four-way directional control valves of this type are frequently used as pilot valves to direct flow to and from another valve (Figure 7-30). Fluid is directed from one source of supply through the rotary valve to another directional control valve, where it positions the valve to direct flow from another source to one side of an actuating unit. Fluid from the other end of the main valve flows through the return line, through the rotary valve, and to the return line.

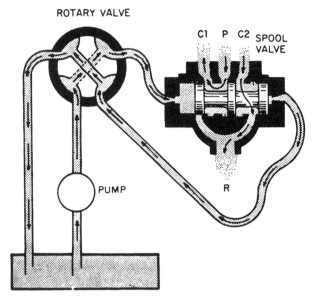

Figure 7–30 Sliding spool valve controlled by a rotary spool valve.

The principal parts of a rotary spool valve were shown in Figure 7-19. Figure 7-31 shows the operation of a rotary spool valve. Views A and C show the valve in a position to deliver fluid to another valve, while view B shows the valve in the neutral position, with all passages through the valve blocked.

Sliding Spool Four-Way Valves

The sliding spool four-way valve is similar in operation to the two-way valve previously described. It is simple in its operation and is the most durable and trouble-free of all four-way control valves.

The valve (Figure 7-32A) consists of a valve body containing four fluid ports: pressure (P), return (R), and two cylinder or work ports (C1 and C2). A hollow sleeve fits into the main bore of the valve body. There are static O-rings placed at intervals around the outside diameter of the sleeve. These O-rings form a seal between the sleeve and the valve body, creating chambers around the sleeve. Each of these chambers is lined up with one of the fluid ports in the body.

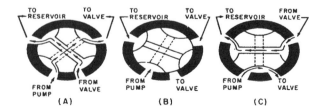

Figure 7–31 Operation of a rotary spool four-way valve.

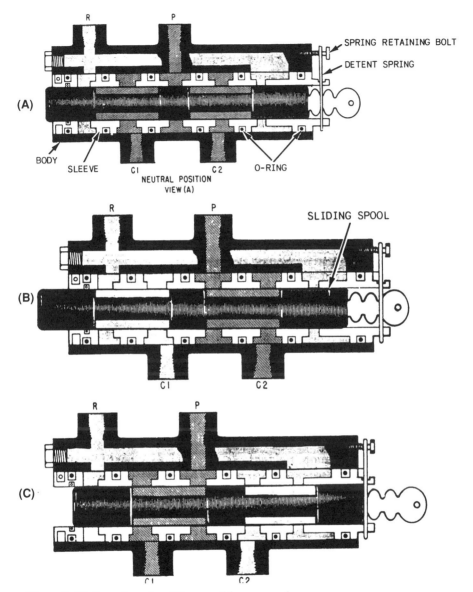

Figure 7–32 Operation of a sliding spool four-way valve.

The drilled passage in the body accounts for a fifth chamber, which results in having two outboard chambers connected to the return port. The sleeve has a pattern of holes drilled through it to allow fluid to flow from one port to another. A series of holes are drilled into the hollow center sleeve in each chamber. The sleeve is prevented from turning by a sleeve retainer bolt or pin that secures it to the valve body.

The sliding spool fits into the hollow center of the sleeve. This spool is similar to the two-way valve, except that this spool has three pistons or lands. These lands are lapped to precisely fit the bore of the sleeve. One or both ends of the spool are connected to actuators attached to the end of the valve body. These actuators can be mechanical, pneumatic, hydraulic, or electrical. When an actuator is activated, it moves the spool to the desired position. The lands of the spool then line up different combinations of fluid ports and permit fluid flow to the selected ports.

The detent spring is a clothespin-type spring, secured to the end of the body by a spring retaining bolt. The two legs of the spring extend down through slots in the sleeve and fit into the detents. The spool is gripped between the two legs of the spring. To move the spool, enough force must be applied to spread the two spring legs and allow them to snap back into the next detent, which would be for another position.

Figure 7-32, view A, shows a manually operated sliding spool valve in the neutral position. The detent spring is in the center detent of the sliding spool. The center land is lined up with the pressure port (P), preventing fluid from flowing into the valve through this port. The return port is also blocked, preventing flow through that port. With both the pressure and return ports blocked, fluid in the actuating line is trapped. For this reason, a relief valve is usually installed in each actuating line when this type of valve is used.

Figure 7-32, view B, shows the valve in the working position with the end of the sliding spool retracted. The detent spring is in the outboard detent, locking the sliding spool in this position. The lands have shifted inside the sleeve and the ports are opened. Fluid, under pressure, enters the sleeve, passes through it by way of the drilled holes, and leaves through cylinder port C2. Return fluid flowing from the actuator enters port C1, then passes through the return port back to the reservoir. Fluid cannot flow past the spool lands because of the lapped surfaces.

Figure 7-32, view C, shows the valve in the opposite working position with the sliding spool extended. The detent spring is in the inboard detent. The center land of the sliding spool is now on the other side of the pressure port, and the fluid, under pressure, is directed through the sleeve and out cylinder port C1. Return fluid flows into the valve through cylinder port C2 and returns to the reservoir.

The directional control valves previously discussed are for use in closed-center fluid power systems. Figure 7-33 shows the operation of a representative open-center sliding spool valve. When this type of valve is in the neutral position (Figure 7-33, view A), fluid flows into the valve through the pressure port (P), through the hollow spool, and returns to the reservoir. When the spool is moved to the right of the neutral position (view B), one working line (C1) is aligned to system pressure and the other working line (C2) is open through the hollow spool to the return port. View C shows the flow of fluid through the valve with the spool moved to the left of neutral position.

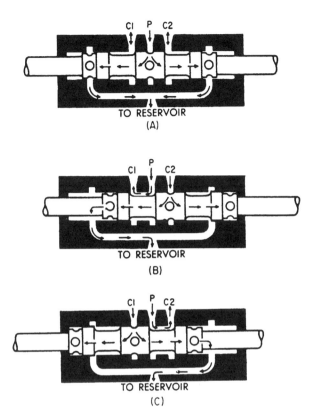

Figure 7–33 Open-center sliding spool valve.

8

LINES, FITTINGS, AND SEALS

The control and application of fluid power would be impossible without suitable means of transferring the hydraulic fluid between the reservoir, the power source, and the points of application. Fluid lines are used to transfer the hydraulic fluid, fittings are used to connect lines to system components, and seals are used in all components to prevent leakage. This chapter is devoted to these critical system components.

TYPES OF LINES

Three types of lines are used in fluid power systems: pipe (rigid), tubing (semirigid), and hoses (flexible). A number of factors are considered when the type of line is selected for a particular application. These factors include the type of fluid, the required system pressure, and the location of the system. For example, heavy pipe might be used for a large, stationary system, but comparatively lightweight tubing must be used in mobile applications. Flexible hose is required in installations where units must be free to move relative to each other.

PIPE AND TUBING

There are three important dimensions of any tubular product: outside diameter (OD), inside diameter (ID), and wall thickness. Sizes of pipe are listed by the nominal, or approximate, ID and the wall thickness. The actual OD and the wall thickness list sizes of tubing.

Selection

The material, inside diameter (ID), and wall thickness are the three primary consider-ations in the selection of lines for a particular fluid power system. The ID of the line is important because it determines how much fluid can pass through the line without loss of power due to excessive friction and heat. The velocity of a given flow is less through a large opening than through a small opening. If the ID of the line is too small for the amount of flow, excessive turbulence and friction will cause unnecessary power loss and overheat the hydraulic fluid.

Sizing

Pipes are available in three different weights: Standard (STD), or Schedule 40; Extra Strong (XS), or Schedule 80; and Double Extra Strong (XXS). The schedule numbers range from 10 to 160 and cover 10 distinct sets of wall thickness (Table 8-1). Schedule 160 wall thickness is slightly thinner than the double extra strong.

As mentioned earlier, the nominal inside diameter (ID) determines the size of pipes. For example, the ID for a 1/4-inch Schedule 40 pipe is 0.364 inch, and the ID for a 1/2-inch Schedule 40 pipe is 0.622 inch.

It is important to note that the IDs of all pipes of the same nominal size are not equal. This is because the OD remains constant and the wall thickness increases as the schedule number increases. For example, a nominal 1-inch Schedule 40 pipe has a 1.049-inch ID. The same size Schedule 80 pipe has a 0.957-inch ID, while Schedule 160 pipe has 0.815-inch ID. In each case the OD is 1.315 inch and the wall thickness varies. The actual wall thickness is the difference between the OD and ID divided by 2.

Tubing differs from pipe in its size classification. Its actual outside diameter (i.e., OD) designates tubing. Thus, 5/8-inch tubing has an OD of 5/8 inch. As indicated in Table 8-2, tubing is available in a variety of wall thicknesses. The diameter of tubing is often measured and indicated in 1/16ths. Thus, No. 6 tubing is 6/16 or 3/8-inch OD, No. 8 tubing is 8/16 or 1/2-inch, and so forth.

The wall thickness, material used, and ID determine the bursting pressure of a line or fitting. The greater the wall thickness in relation to the ID and the stronger the metal,

Table 8–1 Wall Thickness Schedule Designation for Pipe

Nominal size	Pipe OD	Inside diameter (ID)		
		Schedule 40	Schedule 80	Schedule 160
1/8	0.405	0.269	0.215	
1/4	0.540	0.364	0.302	
3/8	0.675	0.493	0.423	
1/2	0.840	0.622	0.546	0.466
3/4	1.050	0.824	0.742	0.815
1	1.315	1.049	0.957	0.815
1-1/4	1.660	1.380	1.278	1.160
1-1/2	1.900	1.610	1.500	1.338
2	2.375	2.067	1.939	1.689

Table 8–2 Tubing Size Designation

Tube OD	Wall thickness	Tube ID	Tube OD	Wall thickness	Tube ID	Tube OD	Wall thickness	Tube ID
1/8	0.028	0.069	5/8	0.035	0.555	1-1/2	0.065 5	1.370
	0.032	0.061		0.042	0.541		0.072	1.356
	0.035	0.055		0.049	0.527		0.083	1.334
				0.058	0.509		0.095	1.310
				0.065	0.495		0.109	1.282
				0.072	0.481		0.120	1.260
				0.083	0.459		0.134	1.232
				0.095	0.435			
3/16	0.032	0.1235	3/4	0.049	0.652	1-3/4	0.065	1.620
	0.035	0.1175		0.058	0.634		0.072	1.606
				0.065	0.620		0.083	1.584
				0.072	0.606		0.095	1.560
				0.083	0.584		0.109	1.532
				0.095	0.560		0.120	1.510
				0.109	0.532		0.134	1.482
1/4	0.035	0.180	7/8	0.049	0.777	2	0.065	1.870
	0.042	0.166		0.058	0.759		0.072	1.856
	0.049	0.152		0.065	0.745		0.083	1.834
	0.058	0.134		0.072	0.731		0.095	1.810
	0.065	0.120		0.083	0.709		0.109	1.782
				0.095	0.685		0.120	1.760
				0.109	0.657		0.134	1.732
5/16	0.035	0.2425	1	0.049	0.902			
	0.042	0.2285		0.058	0.884			
	0.049	0.2145		0.065	0.870			
	0.058	0.1965		0.072	0.856			
	0.065	0.1825		0.083	0.834			
				0.095	0.810			
				0.109	0.782			
				0.120	0.760			
3/8	0.035	0.305	1-1/4	0.049	1.152			
	0.042	0.291		0.058	1.134			
	0.049	0.277		0.065	1.120			
	0.058	0.259		0.072	1.106			
	0.065	0.245		0.083	1.084			
				0.095	1.060			
				0.109	1.032			
				0.120	1.010			
1/2	0.035	0.430						
	0.042	0.416						
	0.049	0.402						
	0.058	0.384						
	0.065	0.370						
	0.072	0.356						
	0.083	0.334						
	0.095	0.310						

the higher the bursting pressure. However, the greater the ID for a given wall thickness, the lower the bursting pressure. This is because force is the product of area and pressure.

Materials

The pipe and tubing used in fluid power systems are commonly made from steel, copper, brass, aluminum, or stainless steel. Each of these metals has its distinct advantages and disadvantages in certain applications.

Steel pipe and tubing are relatively inexpensive and are used in many hydraulic and pneumatic applications. Steel is used because of its strength, suitability for bending and flanging, and adaptability to high pressures and temperatures. Its chief disadvantage is its comparatively low resistance to corrosion.

Copper pipe and tubing are sometimes used for fluid power lines. Copper has high resistance to corrosion and is easily drawn or bent. However, it is unsatisfactory for high temperatures and has a tendency to harden and break because of stress and vibration.

Aluminum has many of the characteristics and qualities required for fluid power lines. It has high resistance to corrosion and is easily drawn or bent. In addition, it has the outstanding characteristic of light weight. Since weight elimination is a vital factor in the design of aircraft, aluminum alloy tubing is used in the majority of aircraft fluid power systems.

An improperly piped system can lead to serious power loss and possible harmful fluid contamination. Therefore, in maintenance and repair of fluid power system lines, the basic design requirements must be kept in mind. Two primary requirements are as follows:

1. The lines must have the correct ID to provide the required volume and velocity of flow with the least amount of turbulence during all demands on the system.
2. The lines must be made of the proper material and have the wall thickness to provide sufficient strength to both contain the fluid at the required pressure and withstand the surges of pressure that may develop in the system.

Preparation

Fluid power systems are designed as compactly as possible, to keep the connecting lines short. Every section of line should be anchored securely in one or more places so that neither the weight of the line nor the effects of vibration are carried on the joints. The aim is to minimize stress throughout the system.

Lines should normally be kept as short and free of bends as possible. However, tubing should not be assembled in a straight line, because a bend tends to eliminate strain by absorbing vibration and also compensates for thermal expansion and contraction.

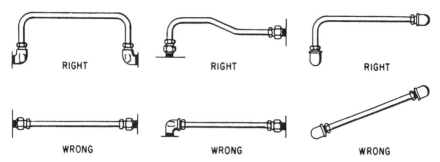

Figure 8–1 Correct and incorrect methods of installation.

Bends are preferred to elbows, because bends cause less loss of power. A few of the correct and incorrect methods of installing tubing are illustrated in Figure 8-1.

Bends are described by their radius measurements. The ideal bend radius is 2-1/2 to 3 times the ID, as shown in Figure 8-2. For example, if the ID of a line is 2 inches, the radius of the bend should be between 5 and 6 inches.

Although friction increases markedly for sharper curves than this, it also tends to increase up to a certain point for gentler curves. The increases in friction in a bend with a radius of more than 3 pipe diameters result from increased turbulence near the outside edges of the flow. Particles of fluid must travel a longer distance in making the change in direction. When the radius of the bend is less than 2-1/2 pipe diameters, the increased pressure loss is due to the abrupt change in the direction of flow, especially for particles near the inside edge of the flow.

Tube Cutting and Deburring

The objective of cutting tubing is to produce a square end that is free from burrs. Tubing may be cut using a standard tube cutter (Figure 8-3) or a fine-toothed hacksaw. When you use the standard tube cutter, place the tube in the cutter with the cutting

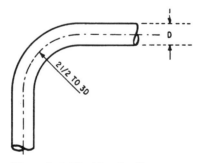

Figure 8–2 Ideal bend radius.

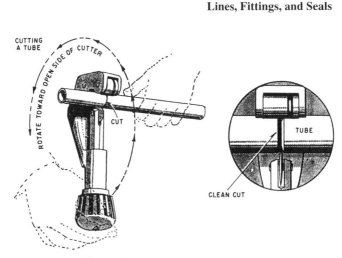

Figure 8–3 Tube cutting.

wheel at the point where the cut is to be made. Apply light pressure on the tube by tightening the adjusting knob. Too much pressure applied to the cutting wheel at one time may deform the tubing or cause excessive burrs. Rotate the cutter, adjusting the tightening knob after each complete rotation.

After the tubing is cut, remove all burrs and sharp edges from the inside and outside of the tube with a deburring tool (Figure 8-4). Clean out the tubing. Make sure no foreign particles remain.

Tube Bending

The objective in tube bending is to obtain a smooth bend without flattening the tube. Tube bending is usually done with either a hand tube bender or a mechanically operated bender.

The hand tube bender (Figure 8-5) consists of a handle, a radius block, a clip, and a slide bar. The handle and slide bars are used as levers to provide the mechanical

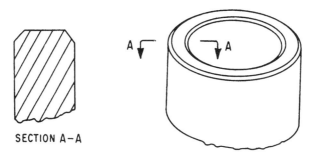

SECTION A–A

Figure 8–4 Properly burred tubing.

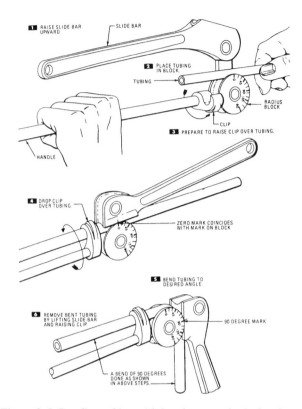

Figure 8–5 Bending tubing with hand-operated tube bender.

advantage necessary to bend the tubing. The radius block is marked in degrees or bend ranging from 0 to 180 degrees. The slide bar has a mark, which is lined up with the zero mark on the radius block. The tube is inserted in the tube bender and after the marks are lined up, the slide bar is moved around until the mark on the slide bar reaches the desired degree of bend on the radius block.

Tube Flaring

Tube flaring is a method of forming the end of a tube into a funnel shape so a threaded fitting can hold it. When a flared tube is prepared, a flare nut is slipped onto the tube and the end of the tube is flared. During tube installation, the flare is seated to a fitting with the inside of the flare against the cone-shaped end of the fitting, pulling the inside of the flare against the seating surface of the fitting.

Either of two flaring tools (Figure 8-6) may be used. One gives a single flare and the other gives a double flare. The flaring tool consists of a split die block that has holes for various sizes of tubing. It also has a clamp to lock the end of the tubing inside the die block and a yoke with a compressor screw and cone that slips over the die block

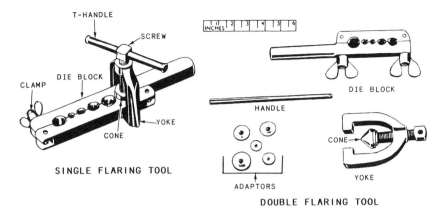

Figure 8–6 Flaring tools.

and forms the 45-degree flare on the end of the tube. A double flaring tool has adapters that turn in the edge of the tube before a regular 45-degree double flare is made.

To use the single flaring tool, first check to see that the end of the tubing has been cut off squarely and has had the burrs removed from both the inside and outside. Slip the flare nut onto the tubing before you make the flare. Then, open the die block. Insert the end of the tubing into the hole corresponding to the OD of the tubing so that the end protrudes slightly above the top face of the die blocks. The amount by which the tubing extends above the blocks determines the finished diameter of the flare. The flare must be large enough to seat properly against the fitting, but small enough that the threads of the flare nut will slide over it. Close the die block and secure the tool with the wing nut. Use the handle of the yoke to tighten the wing nut. Then place the yoke over the end of the tubing and tighten the handle to force the cone into the end of the tubing. The completed flare should be slightly visible above the face of the die blocks.

FLEXIBLE HOSE

Shock-resistant, flexible hose (Figure 8-7) assemblies are required to absorb the movements of mounted equipment under both normal operating conditions and extreme conditions. They are also used for their noise-attenuating properties and to connect moving parts of certain equipment. The two basic hose types are synthetic rubber and poly(tetrafluoroethylene) (PTFE), such as DuPont's Teflon fluorocarbon resin.

Synthetic Rubber Hose

Rubber hoses are designed for specific fluid, temperature, and pressure ranges and are provided in various specifications. Rubber hoses consist of a minimum of three layers: a seamless synthetic rubber tube; one or more reinforcing layers of braided or spiraled

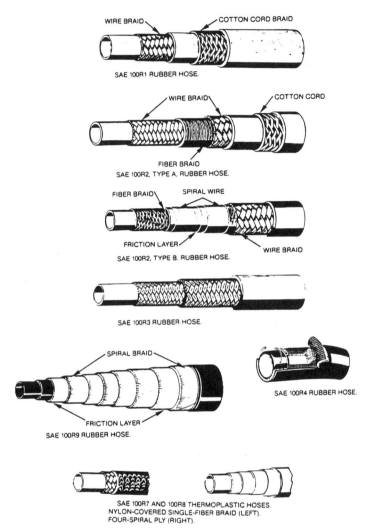

Figure 8–7 Types of flexible hose.

cotton, wire, or synthetic fiber; and an outer cover. The inner tube is designed to withstand the attack of the fluid that passes through it. The braided or spiraled layers determine the strength of the hose. The more layers there are, the greater the pressure rating is. Hoses are provided in three pressure ranges: low, medium, and high. The outer cover is designed to withstand external abuse and contains identification markings.

Sizing

The size of a flexible hose is identified by the dash (–) number, which is the ID of the hose expressed in 16ths of an inch. For example, the ID of a –64 hose is 4 inches. For a few hose styles this is the nominal and not the true ID.

Cure date. Synthetic rubber hoses will deteriorate from aging. A cure date is used to ensure that they do not deteriorate beyond material and performance specifications. The cure date is the quarter and year the hose was manufactured. For example, 1Q89 or 1/89 means the hose was made during the first quarter of 1989. The cure date limits the length of time a rubber hose can be safely used in fluid power applications. The normal shelf life of rubber hose is 4 years.

Application

As mentioned earlier, flexible hose is available in three pressure ranges: low, medium, and high. When replacing hoses, it is important to ensure that the replacement hose is a duplicate of the one removed in length, OD, material, type, and contour. In selecting hose, several precautions must be observed. The selected hose must do all the following:

1. Be compatible with the system fluid
2. Have a rated pressure greater than the design pressure of the system
3. Be designed to give adequate performance and service for infrequent transient pressure peaks up to 150 percent of the working pressure of the hose
4. Have a safety factor with a burst pressure at a minimum of 4 times the rated working pressure

There are temperature restrictions applied to the use of hoses. Rubber hose must not be used where the operating temperature exceeds 200°F. PTFE hoses in high-pressure air systems must not be used where the temperature exceeds 350°F.

Installation

Flexible hose must not be twisted during installation. This will reduce the life of the hose and may cause the fittings to loosen. You can determine whether or not a hose is twisted by looking at the layline that runs along the length of the hose. If the layline does not spiral around the hose, the hose is not twisted. If the layline does spiral around the hose, the hose is twisted and must be untwisted. Flexible hose should be protected from chafing by using a chafe-resistant covering wherever necessary.

The minimum bend radius for flexible hose varies according to the size and construction of the hose and the pressure under which the system operates. Current applicable technical publications contain tables and graphs showing the minimum bend radii for the different types of installations. Bends that are too sharp will reduce the bursting pressure of flexible hose considerably below its rated value.

Flexible hose should be installed so that it will be subjected to a minimum of flexing during operation. Support clamps are not necessary with short installations; but for hose of considerable length (48 inches, for example), clamps should be placed not more than 24 inches apart. Closer supports are desirable and in some cases may be required.

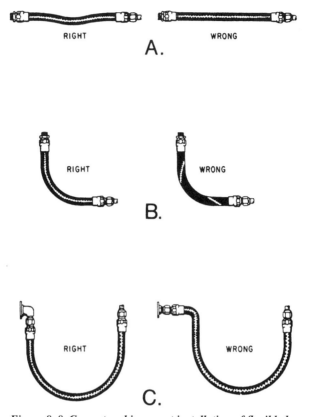

Figure 8–8 Correct and incorrect installation of flexible hose.

A flexible hose must never be stretched tightly between two fittings. About 5 to 8 percent of the total length must be allowed as slack to provide freedom of movement under pressure. When under pressure, flexible hose contracts in length and expands in diameter. Examples of correct and incorrect installations of flexible hose are illustrated in Figure 8-8.

TYPES OF FITTINGS AND CONNECTORS

Some type of connector or fitting must be provided to attach the lines to the components of the system and to connect sections of line to each other. There are many different types of connectors and fittings provided for this purpose. The type of connector or fitting required for a specific system depends on several factors. One determining factor is the type of fluid line (pipe, tubing, or flexible hose) used in the system. Other determining factors are the type of fluid medium and maximum operating pressure of the system. Some of the most common types of fittings and connectors are described in the following sections.

Connectors for Pipes and Tubing

High-pressure pipe or tubing can also be used for hydraulic circuits. In these applications, special threading or fittings are required to connect circuit components.

Threaded Connectors

There are several different types of threaded connectors. In the type discussed in this section, both the connector and the end of the fluid line are threaded. These connectors are used in some low-pressure fluid power systems, are usually made of steel, copper, or brass, and are available in a variety of designs.

Threaded connectors (Figure 8-9) are made with standard pipe threads cut on the inside surface (female). The end of the pipe is threaded with outside threads (male). Standard pipe threads are tapered slightly to ensure tight connections. The amount of taper is approximately 3/4 inch in diameter per foot of thread.

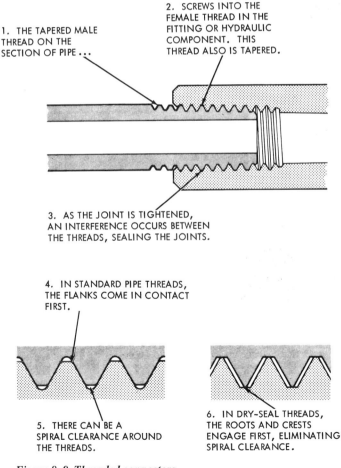

1. THE TAPERED MALE THREAD ON THE SECTION OF PIPE...

2. SCREWS INTO THE FEMALE THREAD IN THE FITTING OR HYDRAULIC COMPONENT. THIS THREAD ALSO IS TAPERED.

3. AS THE JOINT IS TIGHTENED, AN INTERFERENCE OCCURS BETWEEN THE THREADS, SEALING THE JOINTS.

4. IN STANDARD PIPE THREADS, THE FLANKS COME IN CONTACT FIRST.

5. THERE CAN BE A SPIRAL CLEARANCE AROUND THE THREADS.

6. IN DRY-SEAL THREADS, THE ROOTS AND CRESTS ENGAGE FIRST, ELIMINATING SPIRAL CLEARANCE.

Figure 8–9 Threaded connectors.

Metal is removed when pipe is threaded, thinning the pipe and exposing new and rough surfaces. Corrosion agents work more quickly at such points than elsewhere. If pipes are assembled with no protective compounds on the threads, corrosion sets in at once and the two sections stick together so that the threads seize when disassembly is attempted. The result is damaged threads and pipes. To prevent seizing, a suitable pipe thread compound is sometimes applied to the threads. The two end threads must be kept free of compound so that it will not contaminate the hydraulic fluid. Pipe compound, when improperly applied, may get inside the lines and damage pumps, control equipment, and other components of the system.

Another material used on pipe threads is sealant tape. This tape, which is made of Teflon, provides an effective means of sealing pipe connections and eliminates the necessity of torquing connections to excessively high values in order to prevent leaks. It also provides for ease of maintenance whenever it is necessary to disconnect pipe joints. The tape is applied over the male threads, leaving the first thread exposed. After the tape is pressed firmly against the threads, the joint is connected.

Flanged Connectors

Bolted flange connectors (Figure 8-10) are suitable for most pressures now in use. The flanges are attached to the piping by welding, brazing, or tapered threads, or by rolling and bending into recesses. Those illustrated are the most common types of flange joints used. The same types of standard fitting shapes—tee, cross, elbow, and so forth—are manufactured for flange joints. Suitable gasket material must be used between the flanges.

Welded Connectors

Welded joints connect the subassemblies of some fluid power systems, especially in high-pressure systems that use pipe for fluid lines. The welding is done according to standard specifications that define the materials and techniques.

Brazed Connectors

Silver-brazed connectors are commonly used for joining nonferrous piping in the pressure and temperature range where their use is practical. Use of this type of connector is limited to installations in which the piping temperature will not exceed 425°F and the pressure in cold lines will not exceed 3,000 psi. Heating the joint with an oxyacetylene torch melts the alloy. This causes the alloy insert to melt and fill the few thousandths of an inch of annular space between the pipe and fitting.

Flared Connectors

Flared connectors are commonly used in fluid power systems containing lines made of tubing. These connectors provide safe, strong, dependable connections without the need for threading, welding, or soldering the tubing. The connector consists of a fitting, a sleeve, and a nut (Figure 8-11).

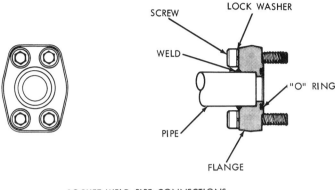

SOCKET WELD PIPE CONNECTIONS
STRAIGHT TYPE

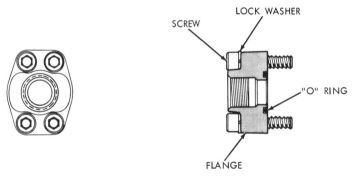

Figure 8–10 Four types of bolted flange connectors.

The fittings are made of steel, aluminum alloy, or bronze. The fitting used in a connection should be made of the same material as the sleeve, the nut, and the tubing. For example, use steel connectors with steel tubing and aluminum alloy connectors with aluminum alloy tubing. Fittings are made in union, 45-degree and 90-degree elbows, tee, and various other shapes (Figure 8-12).

Tees, crosses, and elbows are self-explanatory. Universal and bulkhead fittings can be mounted solidly with one outlet of the fitting extending through a bulkhead and the other outlet(s) positioned at any angle. Universal means the fitting can assume the angle required for the specific installation. Bulkhead means the fitting is long enough to pass through a bulkhead and is designed so it can be secured solidly to the bulkhead.

For connecting to tubing, the ends of the fittings are threaded with straight machine threads to correspond with the female threads of the nut. In some cases, however, one end of the fitting may be threaded with tapered pipe threads to fit threaded ports in

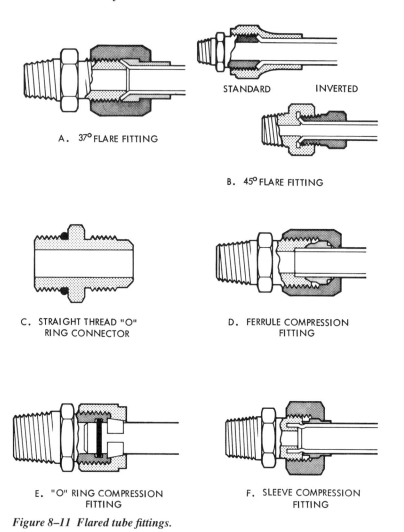

A.   37° FLARE FITTING

STANDARD   INVERTED

B.   45° FLARE FITTING

C.   STRAIGHT THREAD "O"
RING CONNECTOR

D.   FERRULE COMPRESSION
FITTING

E.   "O" RING COMPRESSION
FITTING

F.   SLEEVE COMPRESSION
FITTING

Figure 8–11 Flared tube fittings.

pumps, valves, and other components. Several of these thread combinations are shown in Figure 8-12.

Tubing used with flare connectors must be flared prior to assembly. The nut fits over the sleeve and when tightened, it draws the sleeve and tubing flare tightly against the male fitting to form a positive seal.

The male fitting has a cone-shaped surface with the same angle as the inside of the flare. The sleeve supports the tube so vibration does not concentrate at the edge of the flare and distributes the shearing action over a wider area for added strength.

Correct and incorrect methods of installing flared-tube connectors are illustrated in Figure 8-13. Tubing nuts should be tightened with a torque wrench to the values specified in applicable technical publications.

Figure 8–12 Flared tube fittings.

If an aluminum alloy flared connector leaks after being tightened to the required torque, it must not be tightened further. Overtightening may severely damage or completely cut off the tubing flare, or may result in damage to the sleeve or nut. The leaking connection must be disassembled and the fault corrected.

If a steel tube connection leaks, it may be tightened 1/16 turn beyond the specified torque in an attempt to stop the leakage. If the connection continues to leak, it must be disassembled and the problem corrected.

Connectors for Flexible Hose

There are various types of end fittings for both the piping connection side and the hose connection side of hose fittings. Figure 8-14 shows commonly used fittings.

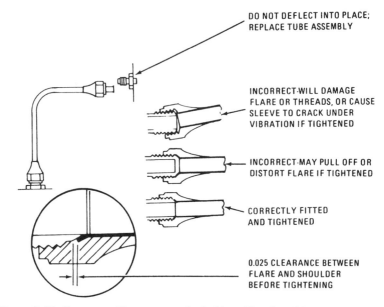

DO NOT DEFLECT INTO PLACE;
REPLACE TUBE ASSEMBLY

INCORRECT-WILL DAMAGE
FLARE OR THREADS, OR CAUSE
SLEEVE TO CRACK UNDER
VIBRATION IF TIGHTENED

INCORRECT-MAY PULL OFF OR
DISTORT FLARE IF TIGHTENED

CORRECTLY FITTED
AND TIGHTENED

0.025 CLEARANCE BETWEEN
FLARE AND SHOULDER
BEFORE TIGHTENING

Figure 8–13 Correct and incorrect method of installing flared fittings.

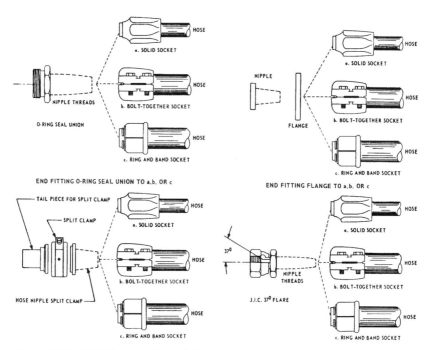

Figure 8–14 End fittings and hose fittings.

Piping Connection Side of Hose Fittings

The piping side of an end fitting comes with several connecting variations: flange, JIC 37° flare, O-ring union, and split clamp, to name a few. Not all varieties are available for each hose.

Hose Connection Side of Hose Fittings

Hose fittings are attached to the hose by several methods. Each method is determined by the fitting manufacturer and takes into consideration such things as size, construction, wall thickness, and pressure rating. Hose used for flexible connections requires one of the following methods for attachment of the fittings to the hose.

> *One-piece reusable sockets.* The socket component of the fitting is fabricated as a single piece. One-piece reusable sockets are screwed or rocked onto the hose OD, followed by insertion of the nipple component.
>
> *Segmented bolted sockets.* The segmented bolted socket consists of two or more segments that are bolted together on the hose after insertion of the nipple component.
>
> *Segmented ring-and-band sockets.* The segmented ring-and-band attachment socket consists of three or more segments. As with the bolted segments, the segments, ring, and band are put on the hose after insertion of the nipple. A special tool is required to compress the segments.
>
> *Segmented ring-and-bolt.* The segmented ring-and-bolt attachment consists of three or more components and is assembled in the same way as the preceding fittings.

QUICK-DISCONNECT COUPLINGS

Self-sealing, quick-disconnect couplings (Figure 8-15) are used at various points in many fluid power systems. These couplings are installed at locations where frequent uncoupling of the lines is required for inspection, test, and maintenance. Quick-disconnect couplings are also commonly used in pneumatic systems to connect sections of air hose and to connect tools to the air pressure lines. This provides a convenient method of attaching and detaching tools and sections of lines without losing pressure.

Quick-disconnect couplings provide a means for quickly disconnecting a line without the loss of fluid from the system or the entrance of foreign matter into the system.

Figure 8–15 Quick-disconnect coupling.

Several types of quick-disconnect couplings have been designed for use in fluid power systems. Figure 8-15 illustrates a coupling that is used with portable pneumatic tools. The male section is connected to the tool or to the line leading from the tool. The female section, which contains the shutoff valve, is installed in the pneumatic line leading from the pressure source. These connectors can be separated or connected by very little effort on the part of the operator.

The most common quick-disconnect coupling for hydraulic systems consists of two parts, held together by a union nut. Each part contains a valve which is held open when the coupling is connected, allowing fluid to flow in either direction. When the coupling is disconnected, a spring in each part closes the valves, preventing the loss of fluid and entrance of foreign matter.

Manifolds

Some fluid power systems are equipped with manifolds in the pressure supply and /or return lines. A manifold is a fluid conductor that provides multiple connection ports. Manifolds eliminate piping, reduce joints, which are often a source of leakage, and conserve space. For example, manifolds may be used in systems that contain several subsystems. One common line connects the pump to the manifold.

There are outlet ports in the manifold to provide connections to each subsystem. A similar manifold may be used in the return system. Lines from the control valves of the subsystem connect to the inlet ports of the manifold, where the fluid combines into one outlet line to the reservoir. Some manifolds are equipped with check valves, relief valves, filters, and so on, that are required for the system. In some cases, the control valves are mounted on the manifold in such a manner that ports of the valves are connected directly to the manifold.

Manifolds are usually one of three types: sandwich, cast, or drilled. The sandwich type is constructed of three or more flat plates. The center plate, or plates, is machined for passages, and the required inlet and outlet ports are drilled into the outer plates. The plates are then bonded together to provide a leakproof assembly. The cast type of manifold is designed with cast passages and drilled ports. The casting may be iron, steel, bronze, or aluminum, depending on the type of system and fluid medium. In the drilled-type manifold, all ports and passages are drilled in a block of metal.

A simple manifold is illustrated in Figure 8-16. This manifold contains one pressure inlet port and several pressure outlet ports that can be blocked off with threaded plugs. This type of manifold can be adapted to systems containing various numbers of subsystems. A thermal relief valve may be incorporated in this manifold. In this case, the port labeled T is connected to the return line to provide a passage for the relieved fluid to flow to the reservoir.

Figure 8-17 shows a flow diagram in a manifold that provides both pressure and return passages. One common line provides pressurized fluid to the manifold, which

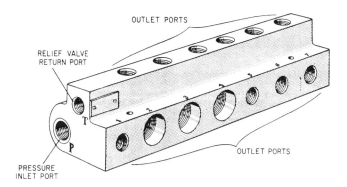

Figure 8–16 Fluid manifold.

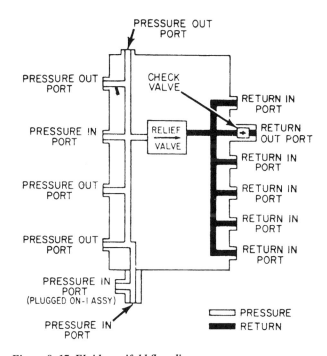

Figure 8–17 Fluid manifold flow diagram.

distributes the fluid to any one of five outlet ports. The return side of the manifold is similar in design. This manifold is provided with a relief valve, which is connected to the pressure and return passages. In the event of excessive pressure, the relief valve opens and allows the fluid to flow from the pressure side of the manifold to the return side.

9

BASIC DIAGRAMS AND SYSTEMS

In the preceding chapters, you have learned about the components that make up a fluid power system. Although a knowledge of system components is essential, it is difficult to understand the interrelationship of these components by simply watching the system operate. Knowledge of system interrelationships is required for effective troubleshooting and maintenance of a fluid power system. Diagrams provided in applicable technical publications or drawings are a valuable aid in understanding the operation of the system and in diagnosing the causes of malfunctions.

This chapter explains the different types of diagrams used to illustrate fluid power circuits, including some of the symbols that depict fluid power components. Included in this section are descriptions and illustrations denoting the differences between open-center and closed-center fluid power systems. The last part of the chapter describes and illustrates some applications of basic fluid power systems.

DIAGRAMS

To troubleshoot a fluid power system, a mechanic or technician must be familiar with the system on which he or she is working. The mechanic must know the function of each component in the system and have a mental picture of its location in relation to other components. Studying the diagrams of the system can provide this knowledge.

A diagram may be defined as a graphic representation of an assembly or system that indicates the various parts and expresses the methods or principles of operations. The ability to read diagrams is a basic requirement for understanding the operation of fluid power systems. Understanding the diagrams of a system requires knowledge of the symbols used in the schematic diagrams.

Symbols

The symbols shown in Appendix A provide a basis for an individual working with fluid power systems to build upon. Some rules applicable to graphical symbols for flow diagrams are as follows:

1. Symbols show connections, flow paths, and the function of the component represented only. They do not indicate conditions occurring during transition from one flow path to another, nor do they indicate component construction or values such as pressure or flow rate.
2. Symbols do not indicate the location of ports, direction of shifting of spools, or position of control elements on actual components.
3. Symbols may be rotated or reversed without altering their meaning except in cases of lines to reservoirs and vented manifolds.
4. Symbols may be drawn to show the normal or neutral condition of each component unless multiple circuit diagrams are furnished showing various phases of circuit operation.

Types of Diagrams

There are many types of diagrams. Those that are most pertinent to fluid power systems are discussed in the following text.

Pictorial Diagrams

Pictorial diagrams (Figure 9-1) show the general location and actual appearance of each component, all interconnecting piping, and the general piping arrangement. This type of diagram is sometimes referred to as an installation diagram.

Diagrams of this type are invaluable to maintenance personnel in identifying and locating components of a system.

Cutaway Diagrams

Cutaway diagrams (Figure 9-2) show the internal working parts of all fluid power components in a system. This includes controls and actuating mechanisms and all interconnecting piping. Cutaway diagrams do not normally use symbols.

Graphic Diagrams

The primary purpose of a graphic or schematic diagram is to enable the maintenance person to trace the flow of fluid from component to component within the system. This type of diagram uses standard symbols to show each component and includes all interconnecting piping. Additionally, the diagram contains a component list, pipe size, data on the sequence of operation, and other pertinent information. The graphic diagram (Figure 9-3) does not indicate the physical location of the various components, but it does show the relation of each component to the other components within the system.

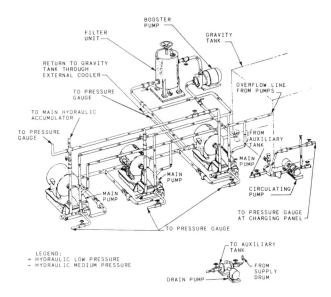

Figure 9–1 Hydraulic system pictorial diagram.

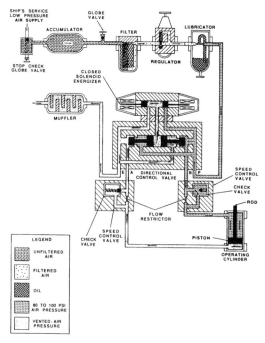

Figure 9–2 Cutaway diagram.

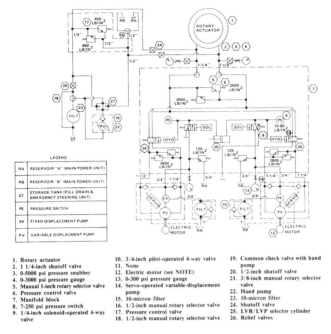

LEGEND

RA	RESERVOIR "A" (MAIN POWER UNIT)
RB	RESERVOIR "B" (MAIN POWER UNIT)
ST	STORAGE TANK (FILL DRAIN & EMERGENCY STEERING UNIT)
PS	PRESSURE SWITCH
PF	FIXED DISPLACEMENT PUMP
PV	VARIABLE DISPLACEMENT PUMP

1. Rotary actuator
2. 1 1/4-inch shutoff valve
3. 0-5000 psi pressure snubber
4. 0-3000 psi pressure gauge
5. Manual 1-inch rotary selector valve
6. Pressure control valve
7. Manifold block
8. 7-250 psi pressure switch
9. 1/4-inch solenoid-operated 4-way valve
10. 3/4-inch pilot-operated 4-way valve
11. None
12. Electric motor (see NOTE)
13. 0-300 psi pressure gauge
14. Servo-operated variable-displacement pump
15. 10-micron filter
16. 1/2-inch manual rotary selector valve
17. Pressure control valve
18. 1/2-inch manual rotary selector valve
19. Common check valve with hand pump
20. 1/2-inch shutoff valve
21. 3/8-inch manual rotary selector valve
22. Hand pump
23. 10-micron filter
24. Shutoff valve
25. LVR/LVP selector cylinder
26. Relief valves

Figure 9–3 Graphic diagram of hydraulic steering unit.

Notice that Figure 9-3 does not indicate the physical location of the individual components with respect to each other in the system. For example, the 3/4-inch solenoid-operated four-way valve (10) is not necessarily located directly above the relief valve (26). The diagram does indicate, however, that the four-way valve is located in the working line between the variable-displacement pump and the 1-inch rotary selector valve and that the valve directs fluid to and from the rotary actuator.

Combination Diagrams

A combination drawing uses a combination of graphic, cutaway, and pictorial symbols. This drawing also includes all interconnecting piping.

FLUID POWER SYSTEMS

A fluid power system in which the fluid in the system remains pressurized from the pump (or regulator) to the directional control valve while the pump is operating is referred to as a closed-center system. In this type of system, any number of subsystems may be incorporated, with a separate directional control valve for each subsystem. The directional control valves are arranged in parallel so that system pressure acts equally on all control valves.

Another type of system that is sometimes used in hydraulically operated equipment is the open-center system. An open-center system has fluid flow by no internal pressure when the actuating mechanisms are idle. The pump circulates the fluid from the reservoir, through the directional control valves, and back to the reservoir (Figure 9-4, view A). Like the closed-center system, the open-center system may have any number of subsystems, with a directional control valve for each subsystem. Unlike the closed-center system, the directional control valves of an open-center system are always connected in series with each other, an arrangement in which the system pressure line goes through each of the directional control valves. Fluid is always allowed free passage through each valve and back to the reservoir until one of the control valves is positioned to operate a mechanism.

When one of the directional control valves is positioned to operate an actuating device, as shown in Figure 9-4, view B, fluid is directed from the pump through one of the working lines to the actuator. With the control valve in this position, the flow of fluid through the valve to the reservoir is blocked. Thus, the pressure builds up in the system and moves the piston of the actuating cylinder. The fluid from the other end of the actuator returns to the control valve through the opposite working line and flows back to the reservoir.

Several different types of directional control valves are used in the open-center system. One type is the manually engaged and manually disengaged directional control valve. After this type of valve is manually moved to the operating position and the actuating mechanism reaches the end of its operating cycle, pump output continues

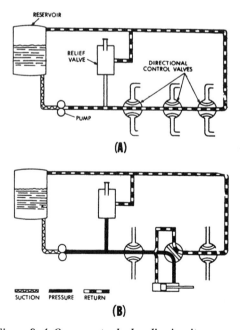

Figure 9–4 Open-center hydraulic circuit.

until the system relief valve setting is reached. The relief valve then unseats and allows the fluid to flow back to the reservoir. The system pressure remains at the pressure setting of the relief valve until the directional control valve is manually returned to the neutral position. This action reopens the open-center flow and allows the system pressure to drop to line resistance pressure.

Another type of open-center directional control valve is manually engaged and pressure disengaged. This type of valve is similar to the valve discussed in the preceding paragraph, but when the actuating mechanism reaches the end of its cycle and the pressure continues to rise to a predetermined pressure, the valve automatically returns to the neutral, open-flow, position.

One advantage of the open-center system is that the continuous pressurization of the system is eliminated. Since the pressure is gradually built up after the directional control valve is moved to an operating position, there is very little shock from pressure surges. This provides a smooth operation of the actuating mechanisms. However, the operation is slower than in a closed-center system, where pressure is always available the moment the directional control valve is positioned. Since most applications require instantaneous operation, closed-center systems are the most widely used.

Hydraulic Power Drive Systems

The hydraulic power system, in its simplest form, consists of the following:

1. The prime mover, which is the outside source of power used to drive the hydraulic pump
2. A variable-displacement hydraulic pump
3. A hydraulic motor
4. A means of introducing a signal to the hydraulic pump to control its output
5. Mechanical shafting and gearing that transmits the output of the hydraulic motor to the equipment being operated

Hydraulic power drives differ in some respects, such as size and method of control. However, the fundamental operating principles are similar. The unit used in the following discussion of fundamental operating principles is representative of hydraulic power drives.

Figure 9-5 shows the basic components of a power drive. The electric motor is constructed with drive shafts at both ends. The forward shaft drives the A-end pump through reduction gears, and the after shaft drives the auxiliary pumps through the auxiliary reduction gears. The reduction gears are installed because the pumps are designed to operate at a speed much slower than that of the motor.

The replenishing pump is a spur gear–type pump. Its purpose is to replenish fluid to the active system of the power drive. It receives its supply of fluid from the reservoir and discharges it to the B-end valve plate. This discharge of fluid from the pump is held at a constant pressure by the action of a pressure relief valve.

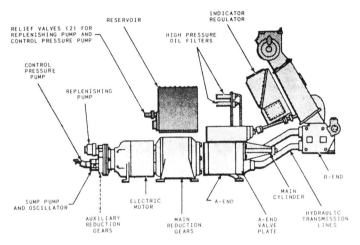

Figure 9–5 Train power drive components.

The sump pump and oscillator has a twofold purpose. It pumps leakage, which collects in the sump of the indicator regulator, to the expansion tank. Additionally, it transmits a pulsating effect to the fluid in the response pressure system. Oscillations in the hydraulic response system help eliminate static friction of valves, allowing hydraulic controls to respond faster.

The control pressure pump supplies high-pressure fluid for the hydraulic control system, brake pistons, lock piston, and the hand-controlled clutch operating piston. The control pressure pump is fixed-displacement, axial-piston type. An adjustable relief valve is used to limit the operating pressure at the outlet of the pump.

Control

For the purpose of this text, control constitutes the relationship between the stroke control shaft and the tilting box. The stroke control shaft is one of the piston rods of a double-acting cylinder. This actuating cylinder and its direct means of control are referred to as the main cylinder assembly (Figure 9-6). It is the link between the hydraulic follow-up system and the power drive itself.

In hand control, the tilting box is mechanically positioned by gearing from the handwheel through the A-end control unit. In local and automatic control, the stroke control shaft positions the tilting box. As shown in Figure 9-6, the extended end of the control shaft is connected to the tilting box. Movement of the shaft will pivot the tilting box one way or the other, which in turn controls the output of the A-end of the transmission. The other end of the shaft is attached to the main piston. A shorter shaft is attached to the opposite side of the piston. This shaft is also smaller in diameter. Thus, the working area of the left side of the piston is twice that of the right side, as it appears in Figure 9-6.

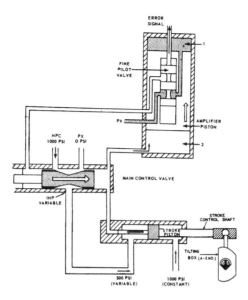

Figure 9–6 Main cylinder assembly.

Intermediate high-pressure (IHP) fluid is transmitted to the left side of the piston, while high-pressure hydraulic (HPC) fluid is transmitted to the right side. The HPC is held constant at 1,000 psi. Since the area of the piston on which HPC acts is exactly one-half the area on which IHP acts, the main piston is maintained in a fixed position, when IHP is one-half (500 psi). Whenever IHP varies from its normal value of 500 psi, the main piston will move, thus moving the tilting box.

Operation

Assume that a right train order signal is received. This will cause the pilot valve to be pulled upward. The fluid in the upper chamber of the amplifier piston can now flow through the lower land chamber of the fine pilot to exhaust. This will cause the amplifier piston to move upward, and the fluid in the right-hand chamber of the main control valve can flow into the lower chamber of the amplifier valve.

The main control valve will now move to the right, IHP will drop below 500 psi, and the stroke piston will move to the left. Movement of the stroke piston will cause tilt to be put on the tilt plate, and the A-end will cause the mount to train right.

Figure 9-7 is a simplified block diagram showing the main element of the hydraulic power drive system under automatic control for clockwise and counterclockwise rotation. There are two principal problems in positioning the unit. One is to get an accurate directional signal. This problem is solved by the director–computer combination. The other problem is to transmit the director signal promptly to the unit so that the position and movements of the unit will be synchronized with the signals from the director.

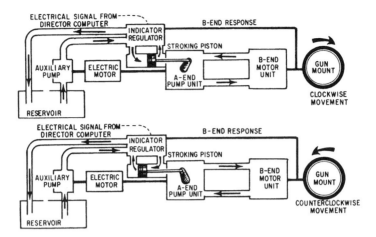

Figure 9–7 Operation of the hydraulic power drive.

The power drive and its control—the indicator regulator—solve the problem of transforming unit movement signals to the hydraulic system. The indicator regulator controls the power drive and this in turn controls movement of the unit.

The indicator regulator receives an initial electrical signal from the director–computer, compares it to the existing unit position, and sends an error signal to the hydraulic control mechanism in the regulator. The hydraulic control mechanism controls flow to the stroke control shaft, which positions the tilting box in the A-end of the transmission. Its tilt controls the volume and direction of fluid pumped to the B-end and, therefore, the speed and direction of the drive shaft of the B-end. Through mechanical linkage the B-end output shaft moves the unit in the direction determined by the input signal. At the same time, B-end response is transmitted to the indicator and continuously combines with incoming director signals to give the error between the two. This error is modified hydraulically, according to the system of mechanical linkages and valves in the regulator. When the unit is lagging behind the signal, its movement is accelerated; and when it begins to catch up, its movement is slowed down so that it will not overrun.

UNLOADING CIRCUITS

An *unloading* circuit is a system where a pump's outlet is diverted to tank at low pressure during part of the cycle. The pump may be unloaded because load conditions at times would exceed the available input power or simply to avoid wasting power and generating heat during idle periods.

Two-Pump Unloading System

It is often desirable to combine the delivery of two pumps for more speed while a cylinder is advancing at low pressure. When the high speed is no longer required or the pressure rises to the point where the combined volume of the two pumps would exceed the input horsepower, the larger of the two pumps is unloaded.

Low-Pressure Operation

Figure 9-8, view A, shows the arrangement of components in such a system and the flow condition at low pressure. Oil from the larger volume pump passes through the unloading valve and over the check valve to combine with the low-volume pump output. This condition continues so long as system pressure is lower than the setting of the unloading valve.

High-Pressure Operation

In view B, system pressure exceeds the setting of the unloading valve, which opens, permitting the large volume pump to discharge to the tank at little or no pressure. The check valve closes, preventing flow from the pressure line through the unloading valve.

In this condition, much less power is used than if both pumps had to be driven at high pressure. However, the final advance of the actuator is slower because of the smaller volume output to the system. When motion stops, the small-volume pump discharges over the relief valve at its unload setting.

Two Maximum Pressures plus Venting

The network shown in Figure 9-9 can be incorporated in a hydraulic system to allow selection of two maximum pressures as well as venting. The highest maximum pressure will be set at the pilot stage of the main relief valve. The remote control relief valve can set a lesser pressure. The solenoid-operated four-way valve switches between the controls.

Venting

In view A of Figure 9-9, both solenoids of the directional control valve are de-energized. The open-center spool is centered by the valve springs, and the vent port on the relief valve is opened to tank. The balanced piston opens and the pump's flow is directed to the tank at the pressure equivalent to the light spring or about 20 psi.

Intermediate Maximum Pressure

In view B of Figure 9-9, the left-hand solenoid of the directional control valve is energized. The valve spool is shifted to connect the relief valve vent port to the remote-controlled valve. This valve now operates as the pilot stage for the balanced piston. Pump flow is diverted to the tank when the remote valve setting is reached.

High Maximum Pressure

In view C of Figure 9-9, the opposite solenoid of the directional control valve is energized. The spool has shifted to connect the relief valve vent port to a *deadheaded*

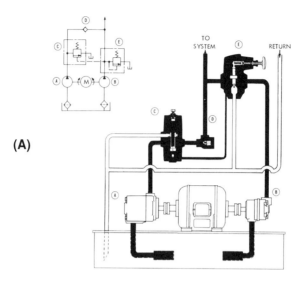

(A)

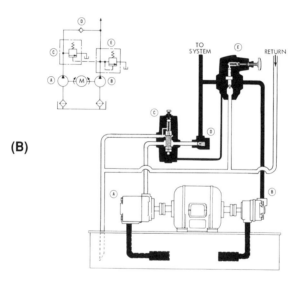

(B)

Figure 9–8 (A) Low-pressure operation; (B) high-pressure operation.

against a plugged port in the directional control valve. The relief valve now functions at the setting of its integral pilot stage.

Automatic Venting at End of Cycle

In systems where it is not necessary to hold pressure at the end of a cycle, it is possible to unload the pump by automatically venting the relief valve. Figure 9-10 shows such a system using a cam-operated pilot valve to vent the relief valve.

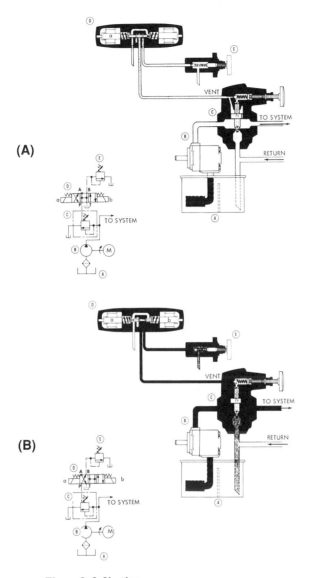

(A)

(B)

Figure 9–9 Venting.

Midstroke Extending

In view A of Figure 9-10, the machine cycle begins when the solenoid of the spring off-set directional valve is energized. Pump output is to the cap end of the cylinder. The vent line from the directional control valve is blocked at the cam-operated pilot valve.

Midstroke Retracting

In view B of Figure 9-10, the limit switch has contracted the cam at the end of the extension stroke. This contact breaks the circuit of the solenoid. The directional

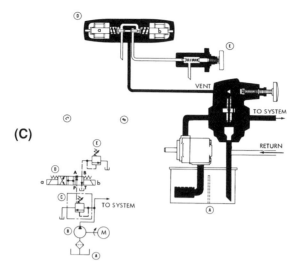

Figure 9–9 Continued.

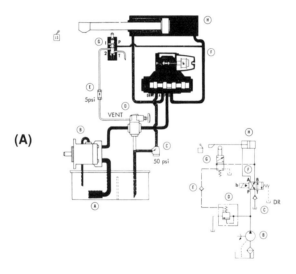

Figure 9–10 Automatic venting.

control valve has shifted to retract the cylinder. The relief valve vent connection is still blocked.

Automatic Stop

View C of Figure 9-10 illustrates the circuit at the end of the retraction stroke. The cam on the cylinder opens the venting pilot valve. The relief valve vent port is thus connected to the line from the cap end of the cylinder and the valve is vented through the inline check

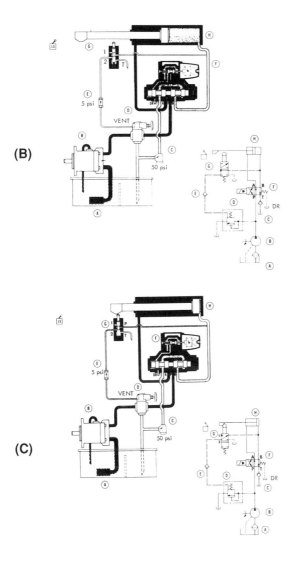

Figure 9–10 Continued.

valve, the directional control valve, and the right-angle check valve. Pilot pressure for the directional valve is maintained at a value determined by the spring loads in the balanced piston of the relief valve, the vent line check valve, and the tank line check valve.

Push-Button Start

When the start button (view D) is depressed, it energizes the solenoid, and the directional valve shifts to direct pump output into the cap end of the cylinder. This causes the check valve in the vent line to close and prevent the relief valve from venting. Pressure again builds up and the cycle is repeated.

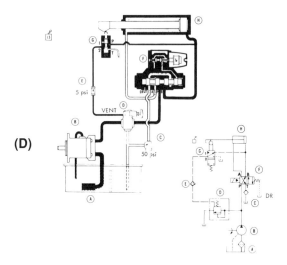

(D)

Figure 9–10 Continued.

ACCUMULATOR SAFETY CIRCUITS

The circuit in Figure 9-11 is used to automatically bleed off a charged accumulator when the pump is shut down to prevent accidental operation of an actuator or to make it safe to open the system for service. The bleed-off is accomplished through a spring-offset directional control valve and a fixed restriction. The directional valve solenoid is actuated by the prime mover switch, so that the solenoid energizes whenever the pump is started (see view A). This blocks the bleed passage during normal operation.

When the pump is shut down (view B), the spool spring shifts the directional control valve and opens the accumulator to tank through the restriction. The manual valve shown is used to control accumulator discharge rate to the system. The auxiliary relief is set slightly higher than the system relief valve and limits pressure rise from heat expansion of the gas charge. The accumulator must have a separator, such as a diaphragm, bladder, or piston, to prevent loss of gas preload each time the machine is shut down.

RECIPROCATING CIRCUITS

Conventional reciprocating circuits use a four-way directional control valve piped directly to a cylinder or hydraulic motor to provide reversal. When a differential cylinder is used, retracting speed is faster than extending speed because of rod volume.

Regenerative Advance

The principle of the regenerative circuit is shown in Figure 9-12. Note that the "B" port on the directional control valve, which would conventionally connect to the cyl-

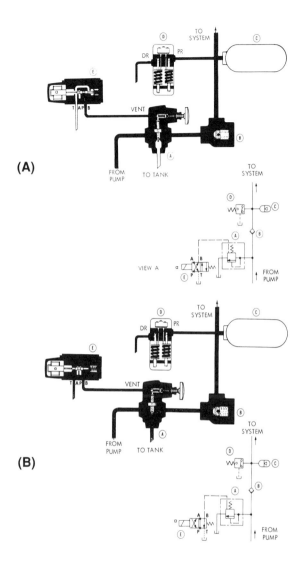

Figure 9–11 Accumulator safety circuit.

inder, is plugged and the rod end of the cylinder is connected directly to the pressure line. With the valve shifted to connect the pressure or "P" port to the cylinder's cap end (view A), flow out of the rod end joins pump delivery to increase the cylinder speed. In the reverse condition (view B), flow from the pump is directly to the rod end of the cylinder. Exhaust flow from the cap end returns to the tank through the directional control valve.

If the ratio of cap end area to rod end annular area in the cylinder is 2:1, the cylinder will advance and retract at the same speed. However, the pressure during advance will be double the pressure required for a conventional circuit. This is because the same

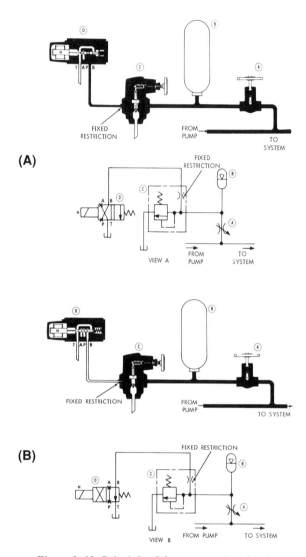

(A)

(B)

Figure 9–12 Principle of the regenerative circuit.

pressure in the rod end, effective over half the cap end area, opposes the cylinder's advance. With a higher ratio of areas, extending speed will increase proportionally.

Regenerative Advance with Conventional Pressure Advance

The regenerative principle also can be used to increase advance speed with a changeover to conventional advance to double the final force. In this circuit (Figure 9-13), a normally closed pressure control valve in effect plugs the "B" port of the directional valve during regenerative advance. When the pressure setting of the pressure

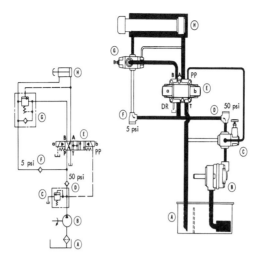

Figure 9–13 Regenerative advance with conventional pressure advance.

control valve is reached, it opens to route oil from the rod end to tank through the directional control valve.

The 5-psi check valve permits oil from the rod end of the cylinder to join pump delivery during regenerative advance, but prevents pump delivery from taking this route to tank during conventional advance. When the directional control valve shifts to retract the cylinder, pump output is through the check valve in the pressure control valve to the rod end of the cylinder.

CLAMPING AND SEQUENCE CIRCUITS

In many applications, such as clamping a work piece and then machining it, it is necessary to have operations occur in a definite order, and to hold pressure at the first operation while the second occurs. Following are two of several such circuits.

Sequencing Circuit

Figure 9-14 shows a method of having machine motions occur in a definite sequence, using one directional control valve and two sequence valves. The counterbalance valve shown is used to control the descent of the vertical cylinder. The sequence is as follows.

> *Cylinder H extends.* Solenoid E-a is energized. This causes the delivery of the hydraulic pump (B) to flow through valves D, E, F1, and the integral check valve of G into the head end of cylinder H. Discharge from the rod end of cylinder H flows freely to the tank through the integral check valve of F2 and valves E and C.

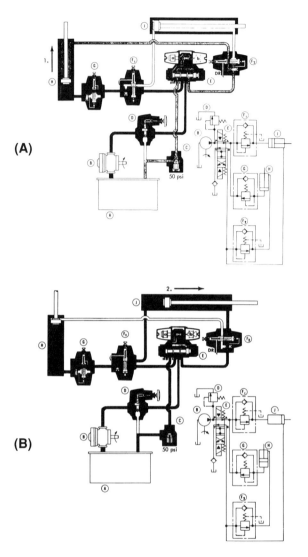

Figure 9–14 Sequencing circuit.

Cylinder J extends, holding pressure on cylinder H. Solenoid E-a remains energized. Pressure increase, on completion of step 1, causes flow to sequence through F1 into the head end of cylinder J. Discharge from the rod end of cylinder J flows freely to tank through valves F2, E, and C. Valve F1 ensures minimum pressure equal to its setting in cylinder H during the extension stroke of cylinder J. When cylinder J is fully extended, pressure increases to the setting of relief valve D, which provides overload protection for the hydraulic pump B.

Cylinder J retracts. Solenoid E-b is energized. Delivery of the hydraulic pump B is directed through valves D, E, and F2 into the rod end of cylin-

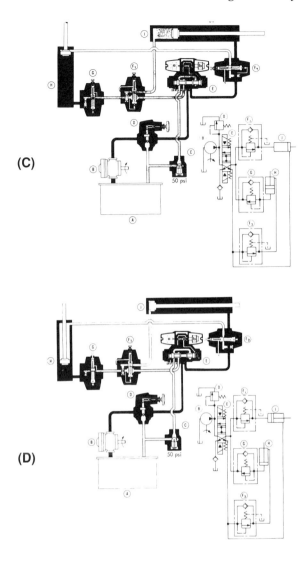

(C)

(D)

Figure 9–14 Continued.

der J. Discharge from cap end of cylinder J flows freely through the integral check valve of F1 and valves E and C to the tank.

Cylinder H retracts. Solenoid E-b remains energized. When cylinder J is fully retracted, pressure increases and causes flow to sequence through F2 into the rod end of cylinder H. Discharge from the head end of cylinder H flows through valve G at its pressure setting and then freely to tank through valves F, E, and C. Valve F2 ensures minimum pressure equal to its setting in the rod end of cylinder J during the retraction of cylinder H. Valve G provides back pressure to prevent cylinder H from falling out of control while lowering.

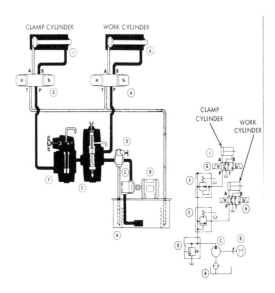

Figure 9–15 Controlled pressure clamping circuit.

Controlled Pressure Clamping Circuit

The circuit shown in Figure 9-15 provides an example of sequencing plus a controlled clamping pressure, which can be held while the work cylinder is feeding and retracting.

Energizing solenoids "b" of valves G and H causes the delivery of hydraulic pump C to be directed through valves D, E , F, and G to the cap end of the clamp cylinder J. The cylinder J extends until the work piece is fully clamped. When the work piece is clamped and pressure builds up to pre-set limits of the sequence valve E, part of the pump's output will be directed from valve E through directional control valve H and into the cap end of work cylinder K, causing it to extend. Valve E ensures minimum pressure, equal to its setting, during operation of cylinder K. Reducing valve F limits the maximum pressure on the cap end of clamping cylinder J.

De-energizing solenoid H-b and energizing solenoid H-a causes delivery of the hydraulic pump C to be diverted through valves E and H to the rod end of the work cylinder K. This causes the work cylinder to fully retract. Solenoid G-b is then de-energized and G-a energized. Delivery from the hydraulic pump is then directed through valves E, F, and G to retract clamp cylinder J.

10

TROUBLESHOOTING HYDRAULIC SYSTEMS

Many of the failures in a hydraulic system show similar symptoms: a gradual or sudden loss of high pressure, resulting in loss of power or speed in the cylinders. In fact, the cylinder may stall under light loads or may not move at all. Often the loss of power is accompanied by an increase in pump noise, especially as the pump tries to build up pressure.

Any major component, i.e., pump, relief valve, directional valve, or cylinder, could be at fault. In a sophisticated system other components could also be at fault, but this would require the services of an experienced technician.

ISOLATING PROBLEMS IN A HYDRAULIC CIRCUIT

By following an organized step-by-step testing procedure in the order given here, the problem can be traced to a general area. Then, if necessary, each component in that area can be tested or replaced.

1. Pump Suction Strainer

Cavitation of the hydraulic pump is the most frequent mode of hydraulic system failure. The most frequent source of cavitation is suction flow restrictions caused by dirt buildup on the suction strainer. This can happen on both new and older systems. It produces the symptoms just described: increased pump noise and loss of high pressure and/or speed.

If the strainer is not located in the pump suction line, it will be found immersed below the oil level in the reservoir. Some operators of hydraulic equipment never give the equipment any attention or maintenance until it fails. Under these conditions, sooner or later, the suction strainer will probably become sufficiently restricted to cause a breakdown of the whole system and damage to the pump.

The suction strainer should be removed for inspection and should be cleaned before reinstallation. Wire mesh strainers can best be cleaned with an air hose, blowing from the inside out. They can also be washed in a solvent that is compatible with the reservoir fluid. Kerosene may be used for strainers operating in petroleum-based hydraulic oil. Do not use gasoline or other explosive or flammable solvents. The strainer should be cleaned even though it may not appear to be dirty. Some clogging materials cannot be seen except by close inspection. If there are holes in the mesh or if there is mechanical damage, the strainer should be replaced.

When reinstalling the strainer, inspect all joints for possible air leaks, particularly at union joints. There must be no air leaks in the suction line. Check reservoir oil level to be sure it covers the top of the strainer by at least 3 inches at minimum oil level, which is with all cylinders extended. If it does not cover to this depth, there is danger of a vortex forming that may allow air to enter the system when the pump is running.

2. Pump and Relief Valve

If cleaning the pump suction does not correct the trouble, isolate the pump and relief valve from the rest of the circuit by disconnecting so that only the pump, relief valve, and pressure gauge remain in the pump circuit. Cap or plug both ends of the plumbing that has been disconnected. The pump is now deadheaded into the relief valve. Start the pump and watch for pressure buildup on the gauge while tightening the adjustment on the relief valve. If full pressure can be developed, obviously the pump and relief valves are operating correctly, and the trouble is to be found further down the line. If full pressure cannot be developed in this test, continue with step 3.

3. Pump or Relief Valve

If high pressure cannot be obtained in step 2 by running the pump against the relief valve, further testing must be conducted to see whether the fault lies in the pump or in the relief valve. Proceed as follows.

If possible, disconnect the reservoir return line from the relief valve. Attach a short length of hose to the relief valve outlet. Hold the open end of this hose over the reservoir filler opening so that the rate of oil flow can be observed. Start the pump and runt the relief valve adjustment up and down while observing the flow through the hose. If the pump is bad, there will probably be a full stream of oil when the relief valve adjustment is backed off, but this flow will diminish or stop as the adjustment is increased. If a flowmeter is available, the flow can be measured and compared with the pump catalog rating.

If a flowmeter is not available, the rate of flow on small pumps can be measured by discharging the hose into a bucket while timing with the sweep hand on a watch. For example, if a volume of 10 gallons is collected in 15 seconds, the pumping rate is 40 gpm.

If the gauge pressure does not rise above a low value, say 100 psi, and if the volume of flow does not substantially decrease as the relief valve adjustment is tightened, the relief valve is probably at fault and should be cleaned or replaced as instructed in step 5.

4. Pump

If a full stream of oil is not obtained in step 3, or if the stream diminishes as the relief valve adjustment is tightened, the pump is probably at fault. Assuming that the suction has already been cleaned and the inlet plumbing has been examined for air leaks, as in step 1, the oil is slipping across the pumping elements inside the pump. This can mean a worn-out pump, or too high an oil temperature. High slippage in the pump will cause the pump to run considerably hotter than the oil reservoir temperature. In normal operation, with a good pump, the pump case will probably run about 200°F above the reservoir temperature. If the temperature difference is greater than this, excess slippage, caused by wear, may be the cause.

Check also for slipping belts, a sheared shaft pin or key, a broken shaft, a broken coupling, or a loosened set screw.

5. Relief Valve

If the test for step 3 has indicated the trouble to be in the relief valve, the quickest remedy is to replace the valve with another one known to be good. The faulty valve may later be disassembled for inspection and cleaning. Pilot-operated relief valves have small orifices that may be blocked with accumulations of dirt. Blow out all passages with an air hose and run a small wire through orifices. Check also for free movement of the spool. In a relief valve with pipe thread connections in the body, the spool may bind if pipe fittings are over-tightened. If possible, test the spool for binding before unscrewing threaded connections from the body, or screw in fittings tightly during inspection of the valve.

6. Cylinders

If the pump will deliver full pressure when operating across the relief valve in step 2, both the pump and relief valve can be considered good. If so, the trouble must be further downstream. The cylinder should be tested first for worn-out or defective packing.

The easiest method for testing a hydraulic cylinder is to run the piston to one end of its stroke and leave it stalled in this position under full pressure. Crack the fitting on the same end of the cylinder to check for fluid leakage. After checking, tighten the fittings and run the piston to the opposite end of its stroke and repeat the test. Occasionally a cylinder will leak at one point in its stroke because of a scratch or dent in the barrel. Check suspected positions in midstroke by installing a positive stop at the suspected position and run the piston rod against it for testing. Once in a great while a piston seal may leak intermittently. This is usually caused by a soft packing or O-ring moving slightly or rolling into different positions on the piston, and is more likely to happen on cylinders of large bore.

When making this test on hydraulic cylinders, the line should be completely removed from a cylinder port during the test and an open line from the valve should be plugged or capped, since a slight back pressure in the tank return would spill oil from the line if not plugged. Pistons with metal ring seals can be expected to have a small amount of leakage across the rings, and even those leak-tight soft seals may have a small bypass during break-in of new seals or after the seals are well worn.

7. Directional Control Valves

If the cylinder has been tested and found to have reasonably tight piston seals, the four-way valve should be checked next. Although it does not often happen, an excessively worn spool can slip enough oil to prevent buildup of maximum pressure. Symptoms of this condition are a loss of cylinder speed together with difficulty in building up to full pressure even with the relief valve adjusted to its highest setting. This condition would be more likely to occur with high-pressure pumps of low-volume output, and it would develop gradually over a long period of time.

For testing four-way valves, it is necessary to obtain access to the tank return ports so that the amount of leakage can be observed. To make the test, disconnect both cylinder lines and plug these ports on the valve. Start up the system and shift the valve to one working position. Any flow out the tank return port while the valve is under pressure is the amount of leakage. Repeat the test in all other working positions of the valve.

FAILURE MODES OF HYDRAULIC COMPONENTS

Each of the components that make up a hydraulic circuit has inherent strengths and weaknesses. These design characteristics define the more common failure modes that may effect each of these components.

Positive-Displacement Pumps

Most hydraulic pumps are positive displacement and are more tolerant to variations in system demands and pressures than centrifugal pumps. However, they are still subject to a variety of common failure modes caused directly or indirectly by the process.

Rotary-Type

Rotary-type, positive-displacement pumps share many common failure modes with centrifugal pumps. Both types of pumps are subject to process-induced failures caused by demands that exceed the pump's capabilities. Operating methods that result in either radical changes in their operating envelope or instability in the process system also cause process-induced failures.

Table 10-1 lists common failure modes for rotary-type, positive-displacement pumps. The most common failure modes of these pumps are generally attributed to problems with the suction supply. They must have a constant volume of clean liquid in order to function properly.

Reciprocating

Table 10-2 lists the common failure modes for reciprocating-type positive-displacement pumps. Reciprocating pumps can generally withstand more abuse and variations in system demand than any other type. However, they must have a consistent supply of relatively clean liquid in order to function properly.

Table 10–1 Common Failure Modes of Rotary-Type Positive-Displacement Pumps

THE CAUSES	No Liquid Delivery	Insufficient Discharge Pressure	Insufficient Capacity	Starts, But Loses Prime	Excessive Wear	Excessive Heat	Excessive Vibration and Noise	Excessive Power Demand	Motor Trips	Elevated Motor Temperature	Elevated Liquid Temperature
Air Leakage Into Suction Piping or Shaft Seal		●	●				●			●	
Excessive Discharge Pressure			●		●		●	●	●		●
Excessive Suction Liquid Temperatures			●	●							
Insufficient Liquid Supply		●	●	●	●		●		●		
Internal Component Wear	●	●	●				●				
Liquid More Viscous Than Design								●	●	●	●
Liquid Vaporizing in Suction Line		●	●	●			●				●
Misaligned Coupling, Belt Drive, Chain Drive					●	●	●	●		●	
Motor or Driver Failure	●										
Pipe Strain on Pump Casing					●	●	●	●		●	
Pump Running Dry	●	●			●	●	●				
Relief Valve Stuck Open or Set Wrong		●	●								
Rotating Element Binding					●	●	●	●	●	●	
Solids or Dirt in Liquid					●						
Speed Too Low		●	●								
Suction Filter or Strainer Clogged	●	●	●				●			●	
Suction Piping Not Immersed in Liquid	●	●		●							
Wrong Direction of Rotation	●	●								●	

The weak links in the reciprocating pump's design are the inlet and discharge valves used to control pumping action. These valves are the most frequent source of failure. In most cases, valve failure is due to fatigue. The only positive way to prevent or minimize these failures is to ensure that proper maintenance is performed regularly on these components. It is important to follow the manufacturer's recommendations for valve maintenance and replacement.

Because of the close tolerances between the pistons and the cylinder walls, reciprocating pumps cannot tolerate contaminated liquid in their suction-supply system. Many of the failure modes associated with this type of pump are caused by contamination (e.g., dirt, grit, and other solids) that enters the suction side of the pump. This problem can be prevented by the use of well-maintained inlet strainers or filters.

Control Valves

Although there are limited common control valve failure modes, the dominant problems are usually related to leakage, speed of operation, or complete valve failure. Table 10-3 lists the more common causes of these failures.

Table 10–2 Common Failure Modes of Reciprocating Positive-Displacement Pumps

THE CAUSES	No Liquid Delivery	Insufficient Capacity	Short Packing Life	Excessive Wear Liquid End	Excessive Wear Power End	Excessive Heat Power End	Excessive Vibration and Noise	Persistent Knocking	Motor Trips
Abrasives or Corrosives in Liquid			●	●					
Broken Valve Springs		●		●			●		
Cylinders Not Filling		●	●	●			●		
Drive-train Problems							●		●
Excessive Suction Lift	●	●							
Gear Drive Problem							●	●	●
Improper Packing Selection			●						
Inadequate Lubrication						●	●		●
Liquid Entry into Power End of Pump						●			
Loose Cross-head Pin or Crank Pin								●	
Loose Piston or Rod								●	
Low Volumetric Efficiency		●	●						
Misalignment of Rod or Packing			●						●
Non-condensables (Air) in Liquid	●	●	●				●		●
Not Enough Suction Pressure	●	●							
Obstructions in Lines	●						●		●
One or More Cylinders Not Operating		●							
Other Mechanical Problems: Wear, Rusted, etc.					●	●	●	●	
Overloading						●			●
Pump Speed Incorrect		●				●			
Pump Valve(s) Stuck Open		●							
Relief or Bypass Valve(s) Leaking		●							
Scored Rod or Plunger		●							●
Supply Tank Empty	●								
Worn Cross-head or Guides			●			●			
Worn Valves, Seats, Liners, Rods, or Plungers	●	●		●					

Special attention should be given to the valve actuator when conducting a Root Cause Failure Analysis. Many of the problems associated with both process and fluid-power control valves are really actuator problems.

In particular, remotely controlled valves that use pneumatic, hydraulic, or electrical actuators are subject to actuator failure. In many cases, these failures are the reason a valve fails to properly open, close, or seal. Even with manually controlled valves, the true root cause can be traced to an actuator problem. For example, when a manually operated process-control valve is jammed open or closed, it may cause failure of the valve mechanism. This overtorquing of the valve's sealing device may cause damage or failure of the seal, or it may freeze the valve stem. Either of these failure modes results in total valve failure.

Table 10–3 Common Failure Modes of Control Valves

	THE CAUSES	Valve Fails To Open	Valve Fails To Close	Leakage Through Valve	Leakage Around Stem	Excessive Pressure Drop	Opens/Closes Too Fast	Open/Closes Too Slow
Manually Actuated	Dirt/Debris Trapped In Valve Seat		●	●				
	Excessive Wear		●	●				
	Galling	●	●					
	Line Pressure Too High	●	●	●	●	●		
	Mechanical Damage	●	●					
	Not Packed Properly				●			
	Packed Box Too Loose				●			
	Packing Too Tight	●	●					
	Threads/Lever Damaged	●	●					
	Valve Stem Bound	●	●					
	Valve Undersized					●		●
Pilot Actuated	Dirt/Debris Trapped In Valve Seat	●	●	●				
	Galling	●	●					
	Mechanical Damage (Seals, Seat)	●	●	●				
	Pilot Port Blocked/Plugged	●	●	●				
	Pilot Pressure Too High		●				●	
	Pilot Pressure Too Low	●		●				●
Solenoid Actuated	Corrosion	●	●	●				
	Dirt/Debris Trapped In Valve Seat	●	●	●				
	Galling	●	●					
	Line Pressure Too High	●	●	●	●			●
	Mechanical Damage	●	●	●				
	Solenoid Failure	●	●					
	Solenoid Wiring Defective	●	●					
	Wrong Type of Valve (N-O, N-C)	●	●					

11

MAINTENANCE OF
HYDRAULIC SYSTEMS

Most companies spend a great deal of money training their maintenance personnel so that they can troubleshoot and correct failures of a hydraulic system. If the focus was shifted to the prevention of system or component failures, less time and money could be spent on troubleshooting. We normally expect hydraulic system failure, rather than deciding not to accept hydraulic failure as the norm. Let's spend the time and money to eliminate hydraulic failure, rather than to prepare for it. I worked for Kendall Company in the 1980s, and we changed our focus from reactive to proactive maintenance on our hydraulic systems, thus eliminating unscheduled hydraulic failure. We will talk about the right way to perform maintenance on a hydraulic system utilizing the *Maintenance Best Practices*.

Lack of maintenance of hydraulic systems is the leading cause of component and system failure, yet most maintenance personnel don't understand the proper maintenance techniques of a hydraulic system. There are two aspects to the basic foundation for proper maintenance of a hydraulic system. The first is preventive maintenance, which is key to the success of any maintenance program, whether for hydraulics or for any equipment of which we require reliability. The second aspect is corrective maintenance, which in many cases can cause additional hydraulic component failure when it is not performed to standard.

PREVENTIVE MAINTENANCE

Preventive maintenance (PM) of a hydraulic system is very basic and simple, and if followed properly it can eliminate most hydraulic component failure. PM is a discipline and must be followed as such in order to obtain results. We must view a PM program as performance-oriented and not activity-oriented. Many organizations have good PM procedures, but do not require maintenance personnel to follow them or hold them accountable for the proper execution of these procedures. In order to develop a preventive maintenance program for your system, you must follow the steps outlined here.

Identify the System Operating Condition

The following questions will help identify the operating condition of the hydraulic system:

Does the system operate 24 hours a day, 7 days a week?

Does the system operate at maximum flow and pressure 70 percent or better during operation?

Is the system located in a dirty or hot environment?

Equipment Manufacturer's Maintenance Requirements

The original equipment manufacturer (OEM) usually provides complete recommendations for the installation, operation, inspection, and preventive maintenance of its hydraulic system. These recommendations should be used as the basis of an effective preventive maintenance program.

System Component Maintenance Requirements

In some cases, hydraulic systems are made up of components that were assembled by the end user. In these cases, a single *Operating and Maintenance* manual will not be available. Therefore, the end user must rely on the manuals provided by each of the component manufacturers. As a minimum, the manuals should include the hydraulic pump, reservoir (Figure 11-1), filters, and each of the control valves and actuators.

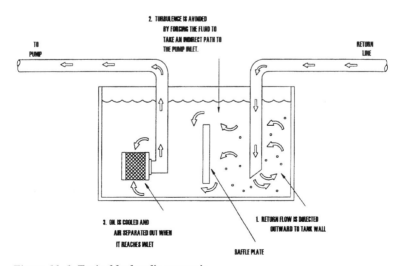

Figure 11–1 Typical hydraulic reservoir.

As in all preventive maintenance programs, we must write procedures required for each PM task. Steps or procedures must be written for each task, and they must be accurate and understandable by all maintenance personnel from entry-level to master.

Preventive maintenance procedures must be a part of the PM Job Plan, which includes the following:

- Tools or special equipment required to perform the task
- Parts or material required to perform the procedure with storeroom number
- Safety precautions for this procedure
- Environmental concerns or potential hazards

A list of preventive maintenance tasks for a hydraulic system could include the following:

1. Change the (could be the return or pressure filter) hydraulic filter.
2. Obtain a hydraulic fluid sample.
3. Filter hydraulic fluid.
4. Check hydraulic actuators.
5. Clean the inside of a hydraulic reservoir.
6. Clean the outside of a hydraulic reservoir.
7. Check and record hydraulic pressures.
8. Check and record pump flow.
9. Check hydraulic hoses, tubing, and fittings.
10. Check and record voltage reading to proportional or servo valves.
11. Check and record vacuum on the suction side of the pump.
12. Check and record amperage on the main pump motor.
13. Check machine cycle time and record.

Preventive Maintenance is the core support that a hydraulic system must have in order to maximize component and life and reduce system failure. Preventive Maintenance procedures that are properly written (Figure 11-2) and followed properly will allow equipment to operate to its full potential and life cycle. Preventive Maintenance allows a maintenance department to control a hydraulic system rather than the system controlling the maintenance department. We must control a hydraulic system by telling it when we will perform maintenance on it and how much money we will spend on the maintenance for the system. Most companies allow hydraulic systems to control their maintenance, at a much higher cost.

In order to validate your preventive maintenance procedures, you must have a good understanding and knowledge of *Best Maintenance Practices* for hydraulic systems. We will discuss these practices now (Table 11-1).

HYDRAULIC KNOWLEDGE

People say that knowledge is power. Well, this is certainly true in hydraulic maintenance. Many maintenance organizations do not know what their maintenance

ABC COMPANY

PREVENTIVE MAINTENANE PROCEDURE

TASK DESCRIPTION:	P.M. – Inspect hydraulic oil reserve tank level
EQUIPMENT NUMBER:	311111
FILE NUMBER:	09
FREQUENCY:	52
KEYWORD, QUALIFIER:	Unit, Hydraulic (Dynamic Press)
SKILL/CRAFT:	Production
PM TYPE:	Inspection
SHUTDOWN REQUIRED:	No

REFERENCE MANUAL/DWGS:
1.   See operator manual F-378

REQUIRED TOOLS/MATERIALS:
1.   Oil, Texaco Rando 68 SDK #400310
2.   Flashlight
3.   Oil Filter Pump

SAFETY PRECAUTIONS:
1.   Observe plant and area specific safe work practices.

MAINTENANCE PROCEDURE:
1.   Inspect hydraulic oil reserve tank level as follows:
a)   If equipped with sight glass, verify oil level at the full mark. Add oil as required.
b)   If not equipped with sight glass, remove fill plug/cap.
c)   Using flashlight, verify that oil is at proper level in tank. Add oil as required.

2.   Record discrepancies or unacceptable conditions in comments.

PM Procedure Courtesy of Life Cycle Engineering, Inc.

Figure 11–2 Example of preventive maintenance task description.

personnel should know. In an industrial maintenance organization, we should divide the hydraulic skill necessary into two groups. In one group are the hydraulic troubleshooters. They must be your experts in maintenance, and as a rule of thumb they should make up 10 percent or less of your maintenance workforce. The other 90+ percent would be your general hydraulic maintenance personnel. They provide the preventive maintenance expertise. The percentages I give are based on a company developing a true preventive/proactive maintenance approach to their hydraulic systems. Let's talk about what knowledge and skills the hydraulic troubleshooter needs.

Hydraulic Troubleshooter: Knowledge

- Mechanical principles: force, work, rate, simple machines
- Math: basic math, complex math equations
- Hydraulic components: application and function of all hydraulic system components
- Hydraulic schematic symbols: meaning of all symbols and their relationship to a hydraulic system
- Calculate flow, pressure, and speed.
- Calculate the system filtration necessary to achieve the system's proper ISO particulate code.

Table 11–1 Best Maintenance Practices for Hydraulic Systems

Component	Component knowledge	Best practices	Frequency
Hydraulic filter	There are two types of filters on a hydraulic system. (1) Pressure filter: Pressure filters come in collapsible and noncollapsible types. Preferred filter is the noncollapsible type. (2) Return filter: Typically has a bypass, which will allow contaminated oil to bypass the filter before indicating the filter needs to be changed.	1. Clean the filter cover or housing with a cleaning agent and clean rags. 2. Remove the old filter with clean hands and install new filter into the filter housing or screw into place. **CAUTION:** NEVER allow your hand to touch a filter cartridge. Open the plastic bag and insert the filter without touching the filter with your hand.	**Preferred:** Based on historical trending of oil samples. **Least preferred:** Based on equipment manufacturer's recommendations.
Reservoir air breather	The typical screen breather should not be used in a contaminated environment. A filtered air breather with a rating of 10 microns is preferred because of the introduction of contaminants to a hydraulic system.	1. Remove and throw away the filter.	**Preferred:** Based on historical trending of oil samples. **Least Preferred:** Based on equipment manufacturer's recommendations.
Hydraulic reservoir	A reservoir is used to: 1. Remove contamination. 2. Dissipate heat from the fluid. 3. Store a volume of oil.	1. Clean the outside of the reservoir to include the area under and around the reservoir. 2. Remove the oil by a filter pump into a clean container, which has not had other types of fluid in it before. 3. Clean the insides of the reservoir by opening the reservoir and cleaning the reservoir with a lint-free rag. 4. Afterwards, spray clean hydraulic fluid into the reservoir and drain out of the system.	If any of the following conditions are met: 1. A hydraulic pump fails. 2. The system has been opened for major work. 3. An oil analysis states excessive contamination.
Hydraulic pumps	Maintenance people need to know the type of pump they have in the system and determine how it operates in their system. Example: What are the flow and pressure of the pump during a given operating cycle? This information allows a maintenance person to trend potential pump failure and troubleshoot a system problem quickly.	5. Check and record flow and pressure during specific operating cycles. 6. Review graphs of pressure and flow. 7. Check for excessive fluctuation of the hydraulic system (designate the fluctuation allowed).	**Pressure checks:** **Preferred:** Daily **Least preferred:** Weekly **Flow and pressure checks:** **Preferred:** Every 2 weeks **Least preferred:** Monthly

Hydraulic Troubleshooter: Skill

- Trace a hydraulic circuit to 100 percent proficiency
- Set the pressure on a pressure-compensated pump
- Tune the voltage on an amplifier card
- Null a servo valve
- Troubleshoot a hydraulic system and utilize root cause failure analysis
- Replace any system component to manufacturer's specification
- Develop a PM Program for a hydraulic system
- Flush a hydraulic system after a major component failure

General Hydraulic: Knowledge

- Filters: function, application, installation techniques
- Reservoirs: function, application
- Basic hydraulic system operation
- Cleaning of hydraulic systems
- Hydraulic lubrication principles
- Proper PM techniques for hydraulics

General Hydraulic: Skills

- Change a hydraulic filter and other system components
- Clean a hydraulic reservoir
- Perform PM on a hydraulic system
- Change a strainer on a hydraulic pump
- Add filtered fluid to a hydraulic system
- Identify potential problems on a hydraulic system
- Change a hydraulic hose, fitting, or tubing

MEASURING SUCCESS

In any program we must track success in order to have support from management and maintenance personnel. We must also understand that any action will have a reaction, which will possibly be negative. We know that successful maintenance programs will provide success but we must have a system of checks and balances to ensure that we're on track.

In order to measure success of a hydraulic maintenance program, we must have some way of tracking success. First, we need to establish a benchmark. A benchmark is a method by which we will establish certain key measurement tools that will tell you the current status of your hydraulic system, and then tell you if you're succeeding in your maintenance program.

Before you begin the implementation of your new hydraulic maintenance program, it would be helpful to identify and track the following information.

Track All Downtime

As in any other type of critical plant system, all delays that affect a hydraulic system's performance should be accurately tracked. The data compiled can then be used to identify inherent design, installation, operation, or maintenance problems that may contribute to its poor reliability. The tracking program should include the following data:

- What component failed?
- Cause of failure?
- Was the problem resolved?
- Could this failure have been prevented?

In addition to tracking the failure mode of hydraulic systems or their components, the program should include an accurate compilation of costs that are incurred as a result of reduced reliability, failures, or poor performance. These costs should include parts and material; labor cost required to repair; lost production; and any other incremental costs that were directly or indirectly caused by the hydraulic system.

Hydraulic Fluid Analysis

Hydraulic fluid becomes contaminated over time or when poor operating and maintenance practices are used. Therefore, an effective preventive maintenance program must include periodic testing of these fluids. In general service applications, representative hydraulic fluid samples should be collected and analyzed on a 30-day interval. Each sample should be tested for the following:

- Copper content
- Silicon content
- Free water
- Iron content
- ISO particulate count
- Fluid condition, i.e., viscosity, additives, and oxidation

When the tracking process begins, you need to trend the information that can be trended (Figure 11-3). These trends let you identify degradation in hydraulic fluid quality that can lead to premature wear, poor performance, or catastrophic failure of critical hydraulic components or systems.

RECOMMENDED MAINTENANCE MODIFICATIONS

Modifications to an existing hydraulic system need to be accomplished professionally. Modifying a hydraulic system in order to improve maintenance efficiency is important to a company's goal of maximum equipment reliability and reduced maintenance cost.

Press Hydraulic System

Hydraulic Fluid Samples

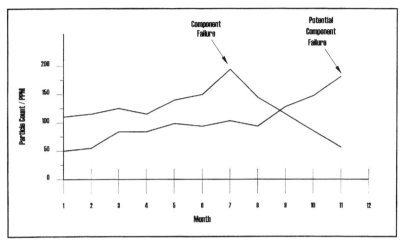

Monthly Samples

Figure 11–3 Hydraulic fluid samples.

1. Filtration Pump with Accessories

Objective. The objective of this pump and modification is to reduce contamination that is introduced into an existing hydraulic system by the addition of new fluid and the device used to add oil to the system (Figure 11-4).

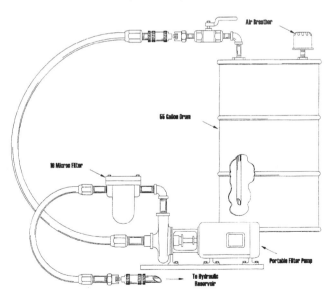

Figure 11–4 Filter pumping unit.

Additional information. Hydraulic fluid from the distributor is usually not filtered to the requirements of an operating hydraulic system. Typically, this oil is strained to a mesh rating and not a micron rating. How clean is clean? Generally, hydraulic fluid must be filtered to 10 microns absolute or less for most hydraulic systems; 25 microns is the size of a white blood cell, and 40 microns is the lower limit of visibility with the unaided eye.

Many maintenance organizations add hydraulic fluid to a system through a contaminated funnel. They may even use a bucket that has contained other types of fluids and lubricants, without cleaning it first.

Recommended equipment and parts.

- Portable filter pump with a filter rating of 3 microns absolute
- Quick disconnects that meet or exceed the flow rating of the portable filter pump.
- A 3/4-inch pipe long enough to reach the bottom of the hydraulic container your fluids are delivered in from the distributor.
- A 2-inch reducer bushing to 3/4-inch npt to fit into the 55-gallon drum, if you receive your fluid by the drum. Otherwise, mount the filter buggy to the double-wall "tote" tank supports, if you receive larger quantities.
- Reservoir vent screens should be replaced with 3/10-micron filters, and openings around piping entering the reservoir should be sealed.

Show a double-wall tote tank of about 300 gallons mounted on a frame for fork truck handling, with the pump mounted on the framework.

Also show pumping from a drum mounted on a frame for fork truck handling, sitting in a catch pan, for secondary containment, with the filter buggy attached.

Regulations require that you have secondary containment, so make everything "leak" into the pan.

2. Modify the Hydraulic Reservoir

Objective. The objective is to eliminate the introduction of contamination through oil being added to the system or contaminates being added through the air intake of the reservoir. A valve needs to be installed for oil sampling (Figure 11-5).

Additional information. The air breather strainer should be replaced with a 10-micron filter if the hydraulic reservoir cycles. A quick disconnect should be installed on the bottom of the hydraulic unit and at the three-quarter level point on the reservoir with valves to isolate the quick disconnects in case of failure. This allows the oil to be added from a filter pump as previously discussed and would allow for external filtering of the hydraulic reservoir oil if needed. Install a petcock valve on the front of the reservoir that will be used for consistent oil sampling.

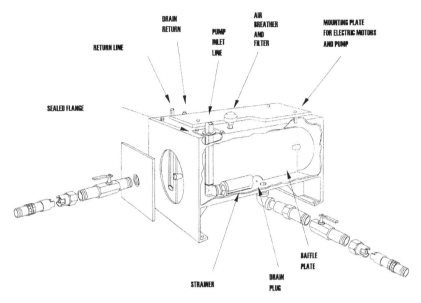

Figure 11–5 Hydraulic reservoir modification.

Equipment and parts needed.

- Quick disconnects that meet or exceed the flow rating of the portable filter pump
- Two gate valves with pipe nipples
- One 10-micron filter breather

WARNING: Do not weld on a hydraulic a reservoir to install the quick disconnects or air filter.

ROOT CAUSE FAILURE ANALYSIS

As in any proactive maintenance organization, you must perform root cause failure analysis in order to eliminate future component failures. Most maintenance problems or failures will repeat themselves unless someone identifies the cause of the failure and proactively eliminates it. A preferred method is to inspect and analyze all component failures. Identify the following:

- Component name and model number
- Location of component at the time of failure
- Sequence or activity the system was performing when the failure occurred
- Cause of the failure
- Method of preventing the failure from happening again

Failures are not caused by an unknown factor such as "bad luck" or "it just happened" or "the manufacturer made a bad part." We have found that most failures can be analyzed and preventive means taken to prevent their reoccurrence. Establishing teams to review each failure can pay off in major ways.

To summarize, maintenance of a hydraulic system is the first line of defense to prevent component failure and thus improve equipment reliability. As we said earlier, discipline is the key to the success of any proactive maintenance program.

Part II

PNEUMATICS

12

PNEUMATIC BASICS

The purpose of pneumatics is to do work in a controlled manner. The control of pneumatic power is accomplished through the use of valves and other control devices that are connected together in an organized circuit. The starting point in this organized circuit is the air compressor, where the air is pressurized.

The pressurized air goes to an air receiver for storage and then is processed for use by passing through filters, dryers, and, in some cases, lubricators. This pressurized air is normally classified as *instrument air* when it is used in control systems. This air must be free of moisture and oil to prevent the control devices from clogging up.

The law states that any pressurized air system must be fitted with a pressure relief valve. This valve prevents the system from being overpressurized and becoming a hazard to personnel or damaging equipment. A pressure switch is an electro-pneumatic control device that is installed on the air receiver to regulate the output of the air compressor. When the air pressure reaches its maximum set point, the regulator is activated and transmits a signal to a solenoid valve on the air compressor. This solenoid valve opens to direct lubricating oil to hydraulically keep the suction valves shut on both the low- and high-pressure cylinders on the air compressor. The air compressor will remain in this mode until the pressure drops to the lower set point and deactivates the pressure regulator. This, in turn, de-energizes the solenoid valve on the air compressor, causing it to release the lubricating oil pressure on the low- and high-pressure suction valves. The air compressor returns to normal operation, pumping air into the receiver until the maximum pressure set point is reached. When this happens, the control cycle starts again. Using a pressure switch prevents the air compressor from running continuously.

Figure 12-1 shows a typical compressed air supply system. The following describes the functions of the system components:

- Compressor. Compresses the air.
- Pressure switch. Turns the air compressor on and off.
- Pressure relief valve. Relieves air pressure at 110 percent of the operating maximum pressure. This device is fitted to the receiver by law.
- Check valve. Permits the compressed air to flow away from the compressor and will not allow any air to return to the compressor.
- Air receiver. The receiver stores the pressurized air. By law, it must have the following fittings installed on it:
  (a) Pressure relief valve
  (b) Pressure gauge
  (c) Access hand-hole
  (d) Drain valve.
- Pressure regulator. Controls the system pressure to the manifold.

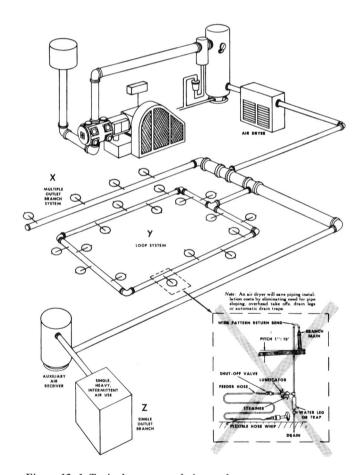

Figure 12–1 Typical compressed air supply.

- Pressure gauge. Indicates internal pressure of the system. Must be fitted by law.
- Filter. This device cleans the air of dirt and contaminants.
- Lubricator. This device adds a small amount of oil to the air in order to lubricate equipment.

Note: This is only installed when needed. Most systems are oil-free.

- Pressure manifold. This distributes the air to the various pressure ports.
- Needle valves. These control the airflow to the various systems that are to be operated.

HAZARDS OF COMPRESSED AIR

People often lack respect for the power in compressed air because air is so common, and it is viewed as harmless. At sufficient pressures, compressed air can cause damage if an accident occurs. To minimize the hazards of working with compressed air, all safety precautions should be followed closely. Reasons for general precautions follow.

Small leaks or breaks in the compressed air system can blow minute particles at surprisingly high speeds. Always wear safety glasses when working in the vicinity of any compressed air system. Goggles in place of glasses are recommended if contact lenses are worn.

Compressors can make an exceptional amount of noise while running. The noise of the compressor, in addition to the drain valves lifting, creates enough noise to require hearing protection. The area around compressors should always be posted as a hearing protection zone.

Pressurized air can do the same type of damage as pressurized water. Treat all operations on compressed air systems with the same care taken with liquid systems. Closed valves should be slowly cracked open and both sides allowed to equalize prior to opening the valve further.

13

CHARACTERISTICS OF COMPRESSED AIR

Pascal's law states that the pressure of a gas or liquid exerts force equally in all directions against the walls of its container. The force is measured in terms of force per unit area (pounds per square inch—psi). This law is for liquids and gases at rest and neglects the weight of the gas or liquid. It should be noted that the field of fluid power is divided into two parts, pneumatics and hydraulics. These two have many characteristics in common. The difference is that hydraulic systems use liquids and pneumatic systems use gases, usually air. Liquids are only slightly compressible and in hydraulic systems this property can often be neglected. Gases, however, are very compressible.

Three properties of gases must be well understood in order to gain an understanding of pneumatic power systems. These are its temperature, pressure, and volume. Compressed air systems and compressors are governed by a number of physical laws that define their efficiency and system dynamics.

THERMODYNAMICS

Both the first and second laws of thermodynamics apply to all compressors and compressed air systems.

First Law of Thermodynamics

This law states that energy cannot be created or destroyed during a process, such as compression and delivery of air or gas, although it may change from one form of energy to another. In other words, whenever a quantity of one kind of energy disappears, an exactly equivalent total of other kinds of energy must be produced.

Second Law of Thermodynamics

This is more abstract and can be stated in several ways:

- Heat cannot, of itself, pass from a colder to a hotter body.
- Heat can be made to go from a body at lower temperature to one at higher temperature only if external work is done.
- The available energy of the isolated system decreases in all real processes.
- Like water, heat or energy will, of itself, flow only downhill.

Basically, these statements say that energy exists at various levels and is available for use only if it can move from a higher to a lower level. In thermodynamics a measure of the unavailability of energy has been devised and is known as *entropy*. It is defined by the differential equation

$$dS = \frac{dQ}{T}$$

Entropy, as a measure of unavailability, increases as a system loses heat, but remains constant when there is no gain or loss of heat, as in an adiabatic process.

BOYLE'S LAW

If a fixed amount of gas is placed in a container of variable volume (such as a cylinder fitted with a piston), the gas will fill completely the entire volume, however large it may be. If the volume is changed, the pressure exerted by the gas will also change. As the volume decreases, the pressure increases. This property is called Boyle's law and can be written as

$$P_1 V_1 = P_2 V_2$$

where P_1 = initial absolute pressure
   V_1 = initial volume of air or gas
   P_2 = final pressure (psia)
   V_2 = final volume of air or gas

According to Boyle's law, the pressure of a gas is inversely proportional to the volume, if the temperature is held constant. For example, 2 cubic feet at 4 psi would exert only 1 psi if allowed to expand to 8 cubic feet.

$$P_1 V_1 = P_2 V_2$$

$$4 \text{ psia} \times 2 \text{ cubic feet} = P_2 \times 8 \text{ cubic feet}$$

$$P_2 = \frac{4 \text{ psia} \times 2 \text{ cubic feet}}{8 \text{ cubic feet}}$$

$$P_2 = 1 \text{ psia}$$

In calculations that involve gas pressure and volume, *absolute pressure*, or pounds per square inch absolute (psia), must be used.

CHARLES' LAW

If a fixed quantity of gas is held at a constant pressure and heated or cooled, its volume will change. According to Charles' law, the volume of a gas at constant pressure is directly proportional to the absolute temperature. This is shown by the following equation:

$$\frac{V_1}{V_2} = \frac{T_1}{T_2}$$

It is important to remember that *absolute temperature* must be considered, not temperature according to normal Fahrenheit or Celsius scales. The absolute Fahrenheit scale is called *Rankin* and the absolute Centigrade is called Kelvin. For conversion:

$$0° \text{ Fahrenheit} = 4,600 \text{ Rankin}$$
$$0° \text{ Celsius} = 2,730 \text{ Kelvin}$$

Thus, gas at 700°F would be 5,300 on the Rankin scale.

COMBINED EFFECT OF PRESSURE, VOLUME, AND TEMPERATURE

Pressure, temperature, and volume are properties of gases that are completely interrelated. Boyle's law and Charles' law may be combined into the following *ideal gas law*, which is true for any gas:

$$\frac{P_1 V_1}{T_1} = \frac{P_2 V_2}{T_2}$$

According to this law, if the three conditions of a gas are known in one situation, then if any condition is changed the effect on the others may be predicted.

DALTON'S LAW

This states that the total pressure of a mixture of ideal gases is equal to the sum of the partial pressures of the constituent gases. The partial pressure is defined as the pressure each gas would exert if it alone occupied the volume of the mixture at the mixture's temperature.

Dalton's law has been proved experimentally to be somewhat inaccurate, the total pressure often being higher than the sum of the partial pressures. This is especially true during transitions as pressure is increased. However, for engineering purposes it is the best rule available and the error is minor.

When all contributing gases are at the same volume and temperature, Dalton's law can be expressed as

$$p = p_a + p_b + p_c + \ldots$$

AMAGAT'S LAW

This is similar to Dalton's law, but states that the volume of a mixture of ideal gases is equal to the sum of the partial volumes that the constituent gases would occupy if each existed alone at the *total* pressure and temperature of the mixture. As a formula this becomes:

$$V = V_a + V_b + V_c + \ldots$$

PERFECT GAS FORMULA

Starting with Charles' and Boyle's laws, it is possible to develop the formula for a given weight of gas:

$$pV = WR^1T$$

where W is weight and R^1 is a specific constant for the gas involved. This is the perfect gas equation. Going one step further, by making W, in pounds, equal to the molecular weight of the gas (1 mole), the formula becomes:

$$pV = R_oT$$

In this very useful form, R_o is known as the *universal gas constant*, has a value of 1,545, and is the same for all gases. The *specific* gas constant (R^1) for any gas can be obtained by dividing 1,545 by the molecular weight. R_o is only equal to 1,545 when gas pressure (p) is in psia; volume (V) is expressed as cubic feet per pound mole; and temperature (T) is in Rankin or absolute, i.e., °F + 460.

AVOGADRO'S LAW

Avogadro's law states that equal volumes of all gases, under the same conditions of pressure and temperature, contain the same number of molecules. This law is very important and is applied in many compressor calculations.

The *mole* is particularly useful when working with gas mixtures. It is based on Avogadro's law that equal volumes of gases at given pressure and temperature (pT) conditions contain equal number of molecules. Because this is so, the *weight* of these equal volumes will be proportional to their molecular weights. The volume of one *mole* at any desired condition can be found by the use of the perfect gas law.

$$pV = R_oT \quad \text{or} \quad pV = 1,545\,T$$

GAS AND VAPOR

By definition, a *gas* is that fluid form of a substance in which it can expand indefinitely and completely fill its container. A *vapor* is a gasified liquid or solid or a substance in gaseous form. These definitions are in general use today.

All gases can be liquefied under suitable pressure and temperature conditions and therefore could also be called vapors. The term gas is most generally used when conditions are such that a return to the liquid state, i.e., condensation, would be difficult within the scope of the operations being conducted. However, a gas under such conditions is actually a superheated vapor.

CHANGES OF STATE

Any given pure substance may exist in three states: as solid, as liquid, or as vapor. Under certain conditions, it may exist as a combination of any two phases, and changes in conditions may alter the proportions of the two phases. There is also a condition where all three phases may exist at the same time. This is known as the *triple point*. Water has a triple point at near 320°F and 14.696 psia. Carbon dioxide may exist as a vapor, a liquid, and a solid simultaneously at about –69.6°F and 75 psia. Under proper conditions, substances may pass directly from a solid to a vapor phase. This is known as *sublimation*.

Changes of State and Vapor Pressure

As liquid physically changes into a gas, its molecules travel with greater velocity and some break out of the liquid to form a vapor above the liquid. These molecules create a *vapor pressure* that, at a specified temperature, is the only pressure at which a pure liquid and its vapor can exist in equilibrium. If in a closed liquid–vapor system the volume is reduced at constant temperature, the pressure will increase imperceptibly until condensation of part of the vapor into liquid has lowered the pressure to the original vapor pressure corresponding to the temperature. Conversely, increasing the volume at constant temperature will reduce the pressure imperceptibly and molecules will move from the liquid phase to the vapor phase until the original vapor pressure has been restored. For every substance, there is a definite vapor pressure corresponding to each temperature.

The temperature corresponding to any given vapor pressure is obviously the *boiling point* of the liquid and also the *dew point* of the vapor. Addition of heat will cause the liquid to boil and removal of heat will start condensation. The three terms—saturation temperature, boiling point, and dew point—all indicate the same physical temperature at a given vapor pressure. Their use depends on the context in which they appear.

Critical Gas Conditions

There is one temperature above which a gas cannot be liquefied by pressure increase. This point is called the *critical temperature*. The pressure required to compress and condense a gas at this critical temperature is called the *critical pressure*.

RELATIVE HUMIDITY

Relative humidity is a term frequently used to represent the quantity of moisture or water vapor present in a mixture although it uses partial pressures in so doing. It is expressed as:

$$RH(\%) = \frac{\text{Actual partial vapor pressure} \times 100}{\text{Saturated vapor pressure at existing mixture temperature}}$$

$$= \frac{p_v \times 100}{p_\lambda}$$

Relative humidity is usually considered only in connection with atmospheric air, but since it is unconcerned with the nature of any other components or the total mixture pressure, the term is applicable to vapor content in any problem. The saturated water vapor pressure at a given temperature is always known from steam tables or charts. It is the existing partial vapor pressure, that is desired and therefore calculable when the relative humidity is stated.

SPECIFIC HUMIDITY

Specific humidity, used in calculations on certain types of compressors, is a totally different term. It is the ratio of the weight of water vapor to the weight of *dry air* and is usually expressed as pounds, or grains, of moisture per pound of dry air. Where p_a is the partial air pressure, specific humidity can be calculated as:

$$SH = \frac{W_v}{W_a}$$

$$SH = \frac{0.662 p_v}{p p_v} = \frac{0.662 p_v}{p_a}$$

DEGREE OF SATURATION

The degree of saturation denotes the actual relationship between the weight of moisture existing in a space and the weight that would exist if the space were saturated:

$$\text{Degree of Saturation }(\%) = \frac{SH_{\text{actual}} \times 100}{SH_{\text{saturated}}}$$

A great many dynamic compressors handle air. Their performance is sensitive to density of the air, which varies with moisture content. The practical application of partial pressures in compression problems centers to a large degree on the determination of mixture volumes or weights to be handled at the intake of each stage of compression, the determination of mixture molecular weight, specific gravity, and the proportional or actual weight of components.

PSYCHROMETRY

Psychrometry has to do with the properties of the air–water vapor mixtures found in the atmosphere. Psychrometry tables, published by the U.S. Weather Bureau, give detailed data about vapor pressure, relative humidity, and dew point at the sea-level barometric pressure of 30 in Hg, and at certain other barometric pressures. These tables are based on relative readings of dry bulb and wet bulb atmospheric temperatures as determined simultaneously by a sling psychrometer. The dry bulb reads ambient temperature while the wet bulb reads a lower temperature influenced by evaporation from a wetted wick surrounding the bulb of a parallel thermometer.

COMPRESSIBILITY

All gases deviate from the perfect or ideal gas laws to some degree. In some cases the deviation is rather extreme. It is necessary that these deviations be taken into account in many compressor calculations to prevent compressor and driver sizes being greatly in error.

Compressibility is experimentally derived from data about the actual behavior of a particular gas under pVT changes. The compressibility factor, Z, is a multiplier in the basic formula. It is the ratio of the actual volume at a given pT condition to ideal volume at the same pT condition. The ideal gas equation is therefore modified to:

$$pV = ZR_oT \text{ or } Z = \frac{pV}{R_oT}$$

In this equation, R_o is 1,545 and p is pounds per square foot.

GENERATION OF PRESSURE

Keeping with the subject of pressure, the basic concepts will be treated in the working sequence pressure generation, transmission, storage, and utilization in a pneumatic system.

Pumping quantities of atmospheric air into a tank or other pressure vessel produces pressure. Pressure is *increased* by progressively increasing the amount of air in a confined space. The effects of pressure exerted by a confined gas result from the average of forces acting on container walls caused by the rapid and repeated bombardment from an enormous number of molecules present in a given quantity of air. This is accomplished in a controlled manner by *compression*, a decrease in the space between the molecules. Less volume means that each particle has a shorter distance to travel; thus, proportionately more collisions occur in a given span of time, resulting in a higher pressure. Air compressors are designed to generate particular pressures to meet individual application requirements.

Basic concepts discussed here are atmospheric pressure; vacuum; gauge pressure; absolute pressure; Boyle's law or pressure/volume relationship; Charles' law or temperature/volume relationship; combined effects of pressure, temperature, and volume; and generation of pressure or compression.

Atmospheric Pressure

In the physical sciences, pressure is usually defined as the perpendicular force per unit area, or the stress at a point within a confined fluid. This force per unit area acting on a surface is usually expressed in pounds per square inch.

The weight of the earth's atmosphere pushing down on each unit of surface constitutes atmospheric pressure, which is 14.7 psi at sea level. This amount of pressure is called *one atmosphere*. Because the atmosphere is not evenly distributed about the earth, atmospheric pressure can vary, depending upon geographic location. Also, obviously, atmospheric pressure decreases with higher altitude. A barometer using the height of a column of mercury or other suitable liquid measures atmospheric pressure.

Vacuum

It is helpful to understand the relationship of vacuum to the other pressure measurements. Vacuums can range from atmospheric pressure down to "zero absolute pressure," representing a "perfect" vacuum (a theoretical condition involving the total removal of all gas molecules from a given volume). The amount of vacuum is measured with a device called a vacuum gauge.

Vacuum is a type of pressure. A gas is said to be under vacuum when its pressure is below atmospheric pressure, i.e., 14.7 psig at sea level. There are two methods of stating this pressure, but only one is accurate in itself.

A differential gauge that shows the difference in the system and the atmospheric pressure surrounding the system usually measures vacuum. This measurement is expressed in the following units:

Millimeters of mercury—vacuum	(mm Hg Vac)
Inches of mercury—vacuum	(in Hg Vac)
Inches of water—vacuum	(in H_2O Vac)
Pounds per square inch—vacuum	(psi Vac)

Unless the barometric or atmospheric pressure is also given, these expressions do not give an accurate specification of pressure. Subtracting the *vacuum* reading from the atmospheric pressure will give an absolute pressure, which is accurate. This may be expressed in the following units:

Inches of mercury—absolute	(in Hg Abs)
Millimeters of mercury—absolute	(mm Hg Abs)
Pounds per square inch—absolute	(psia)

The word *absolute* should never be omitted; otherwise one is never sure whether a vacuum is expressed in differential or absolute terms.

Perfect Vacuum

A perfect vacuum is space devoid of matter. It is absolute emptiness. The space is at zero pressure *absolute*. A perfect vacuum cannot be obtained by any known means, but can be closely approached in certain applications.

Gauge Pressure

Gauge pressure is the most often used method of measuring pneumatic pressure. It is the relative pressure of the compressed air within a system. Gauge pressure can be either positive or negative, depending upon whether its level is above or below the atmospheric pressure reference. Atmospheric pressure serves as the *reference level* for the most significant types of pressure measurements. For example, if we inflate a tire to 30 psi, an ordinary tire-pressure gauge will express this pressure as the value in *excess* of atmospheric pressure, or 30 psig ("g" indicates gauge pressure). This reading shows the numerical value of the *difference* between atmospheric pressure and the air pressure in the tire.

Absolute Pressure

A different reference level, absolute pressure, is used to obtain the total pressure value. Absolute pressure is the *total* pressure, i.e., gauge and atmospheric, and is expressed as psia or pounds per square inch—absolute. To obtain absolute pressure, simply add the value of atmospheric pressure (14.7 psi at sea level) to the gauge pressure reading.

Absolute pressure (psia) values must be used when computing the pressure changes in a volume and when pressure is given as one of the conditions defining the amount of gas contained within a sample.

14

COMPRESSORS

COMMON COMPONENTS

A compressor must operate within a system that is designed to acquire and compress a gas. These systems must include certain components, regardless of compressor type.

Lubrication System

The lubrication system has two basic functions: to lubricate the compressor's moving components and to cool the system by removing heat from the compressor's moving parts. Although all compressors must have a lubrication system, the actual design and function of these systems will vary depending on compressor type.

Centrifugal Compressors

The lubricating system for centrifugal or dynamic compressors is designed to provide bearing lubrication. In smaller compressors, the lubrication systems may consist of individual oil baths located at each of the main shaft bearings. In larger compressors, such as a bullgear design, a positive system is provided to inject oil into the internal, tilting-pad bearings located at each of the pinion shafts inside the main compressor housing.

In positive lubrication systems, a gear-type pump is normally used to provide positive circulation of clean oil within the compressor. In some cases, the main compressor shaft directly drives this pump. In others, a separate motor-driven pump is used.

Positive-Displacement Compressors

Positive-displacement compressors use their lubrication system to provide additional functions. The lubrication system must inject sufficient quantities of clean fluid to provide lubrication for the compressor's internal parts, such as pistons and lobes, and to provide a positive seal between moving and stationary parts.

The main components of a positive-displacement compressor's lubrication system consist of an oil pump, filter, and heat exchanger. The crankcase of the compressor acts as the oil sump. A lockable drain cock is installed at the lowest end of the crankcase to permit removal of any water accumulation that has resulted from sweating of the crankcase walls. The oil passes through a strainer into the pump. It then flows through the heat exchanger, where it is cooled. After the heat exchanger, the cooled oil flows directly to the moving parts of the compressor before returning to the crankcase sump. A small portion is diverted to the oil injector, if one is installed. The oil that is injected into the cylinder seals the space between the cylinder wall and the piston rings. This prevents compressed air from leaking past the pistons, and thus improves the compressor's overall efficiency.

Lube Pump

The oil pump is usually gear driven from the crankshaft so that it will start pumping oil immediately on startup of the compressor. In compressors that work in an oil-free system, oil injectors are not used. Oil separators are installed on the discharge side after leaving the aftercooler.

Oil Separator

The basic purpose of an oil separator is to clean the pressurized air of any oil contamination, which is highly detrimental to pneumatically controlled instrumentation. A separator consists of an inlet, a series of internal baffle plates, a wire mesh screen, a sump, and an outlet (Figure 14-1). The pressurized air enters the separator and immediately passes through the baffle plates. As the air impinges on the baffle plates, it is forced into making sharp directional changes as it passes through each baffle section. As a result, the oil droplets separate from the air and collect on the baffles before dropping into the separator's sump.

After the air clears the baffle section, it then passes through the wire mesh screen where any remaining oil is trapped.

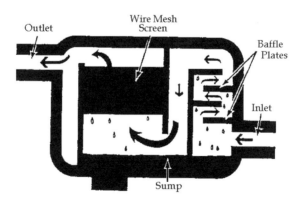

Figure 14–1 Basic oil separator system.

The relatively oil-free air continues to the air reservoir for storage. The air reservoir acts as a final separator where moisture and oil is eventually removed. The air reservoir has drain traps installed at its lowest point where any accumulated moisture/oil is automatically discharged.

As a part of any routine maintenance procedure, these discharge traps should periodically be manually bypassed to ensure that the trap is functioning, and no excessive water accumulation is evident.

POSITIVE-DISPLACEMENT COMPRESSORS

As their name implies, this classification of compressor will displace a constant volume of gas for each complete cycle or rotation of its primary shaft. Unlike centrifugal compressors, a positive-displacement compressor will continue to generate pressure until it exceeds its maximum limits. This design feature can create problems in compressed-air systems. Systems designed to operate at 100 psig may reach line pressures of 150 psig or more. Therefore, it is imperative that all positive-displacement systems include a positive means of relieving excess pressure.

Normally, compressed air systems that use positive-displacement compressors will include both pressure relief valves and an automatic pressure control system that periodically unloads the compressor.

There are three basic types of positive-displacement compressors: reciprocating, rotary screw, and vane type. Although the actual configuration and dynamics of these compressors vary, they all share common design concepts.

RECIPROCATING COMPRESSORS

Reciprocating compressors are the most widely used of all compression equipment and also provide the widest range of sizes and types. Ratings vary from fractional to more than 12,000 horsepower units. Pressures range from low vacuums, at intake, to special process compressors for 60,000 psig or higher.

In common with all positive-displacement compressors, the reciprocating type is classified as a *constant-volume variable-pressure* machine. They are the most efficient type of compressor and can be used in partial loads or reduced capacity applications. Because of the reciprocating pistons and other parts, as well as some unbalanced rotating parts, inertia forces are set up that tend to shake the unit. It is necessary to provide a mounting that will stabilize the installation. The extent of this requirement will depend on the type and size of the compressor.

Reciprocating compressors should be supplied with clean gas. Inlet filters are recommended in all applications. They cannot satisfactorily handle liquids that may be entrained in the gas, although vapors are no problem if condensation within the cylinders

does not take place. Liquids tend to destroy lubrication and cause excessive wear. Reciprocating compressors deliver a pulsating flow of gas. This is sometimes a disadvantage,
but pulsation dampers can usually eliminate the problem.

Design Fundamentals

There are certain design fundamentals that should be clearly understood before
attempting to analyze the operating condition of reciprocating compressors. The fundamentals include the following.

Frame and Running Gear

Two basic factors guide the designer of the frame and running gear. First is the maximum horsepower to be transmitted through the shaft and running gear to the cylinder
pistons; second is the load imposed on the frame parts by the pressure differential
between the two sides of each piston. The latter is often called *pin load* because this
full force is directly exerted on the crosshead and crankpin. These two factors determine the size of bearings, connecting rods, frame, and bolts that must be used
throughout the compressor and in its support structure.

Cylinder Design

The efficiency of compression is entirely dependent upon the design of the cylinder
and its valves. Unless the valve area is sufficient to allow gas to enter and leave the
cylinder without undue restriction, efficiency cannot be high. Valve placement for free
flow of gas in and out of the cylinder is also important.

The method of cylinder cooling must be consistent with the service intended, since
both efficiency and maintenance are influenced by the degree of cooling during compression. The cylinders and all its parts must be designed to withstand the maximum
pressure to which it will be exposed, using those materials that will economically give
the proper strength and the longest service under design conditions.

Inlet and Discharge Valves

Compressor valves are devices placed in each cylinder to permit one-way flow of gas
either into or out of the cylinder. There must be one or more for inlet and discharge in
each compression chamber.

Each valve must open and close once for each revolution of the crankshaft. The valves
in a compressor operating at 700 rpm for 8 hours per day and 250 days per year will
have opened and closed 42,000 times per hour, 336,000 times per day, or 84 million
times in a year.

They have less than 1/10 of a second to open, let the gas pass through, and close. They
must do this with a minimum of resistance or power will be wasted. They must have
small clearance space or there will be excessive re-expansion during the suction
stroke and reduced volumetric efficiency. They must be tight under severe pressure
and temperature conditions. Finally, they must be durable under many kinds of abuse.

Types of Valves

There are four basic valve designs used in various compressors. These are finger, channel, leaf, and annular ring. Within each class there may be variations in design details depending upon operating speed and size of valve required.

Finger. Figure 14-2 shows a typical finger valve before assembly. These valves are used for smaller, air-cooled compressors. One end of the finger is fixed and the opposite end lifts when the valve opens.

Channel. The channel valve (Figure 14-3) is widely used in mid-sized to large compressors. This type of valve uses a series of separate stainless steel channels as valves. This is a cushioned valve as explained in the figure, which adds greatly to its life.

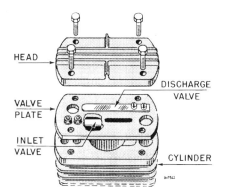

Figure 14–2 Finger valve configuration.

Figure 14–3 Channel valve configuration.

Leaf. The leaf valve has a configuration somewhat like the channel. It is flat strip steel that opens against an arched stop plate so that the valve flexes with maximum lift only at its center. The valve (Figure 14-4) is its own spring.

Annular ring. Figure 14-5 illustrates both inlet and discharge explosion drawings of a typical annular ring valve. The example has a single ring. Larger sizes may have two or three rings. Some designs have the concentric rings tied into a single piece by bridges. The valve shown was the first cushioned valve built.

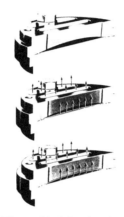

Figure 14–4 Leaf spring configuration.

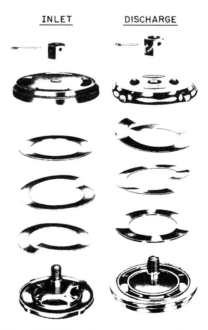

Figure 14–5 Annular ring valve configuration.

The springs and valve move into a recess in the stop plate as the valve opens. Gas trapped in the recess acts as a cushion and prevents slamming. This eliminates a major source of valve and spring breakage.

Cylinder Cooling

Heat in a cylinder is produced by the work of compression, plus the friction of the piston and piston rings on the cylinder wall and of the rod packing on the rod. Heat can be considerable, particularly when moderate and high compression ratios are involved. Undesirably high operating temperatures can be developed.

Most compressors use some method of dissipating a portion of this heat, thus reducing both cylinder wall temperature and final gas temperature. There are several advantages in cylinder cooling, at least, some of which apply to all but exceptional cases.

Lowering cylinder wall and cylinder head temperature reduces losses in capacity and horsepower per unit volume due to suction gas preheating during the inlet stroke. There will be a greater weight of gas in the cylinder ready to be compressed.

Reducing cylinder wall and cylinder head temperature will remove more heat from the gas during compression, lowering its final temperature and reducing power required.

A reduction in gas temperature and in that of the metal surrounding the valves provides a better operating climate for these parts, giving longer valve service life and reducing the possibility of deposit formation.

Reduced cylinder wall temperature promotes better lubrication, resulting in longer life and reduced maintenance

Cooling, particularly water cooling, maintains a more even temperature around the cylinder bore and reduces warpage.

Cylinder Orientation

Orientation of the cylinders in a multistage or multicylinder compressor has a direct effect on its operating dynamics and the vibration level generated by the compressor.

Figure 14-6 illustrates a typical three-piston, air-cooled compressor. Since there are three pistons oriented within a 90-degree arc, this type of compressor will generate higher levels of vibration than an opposed piston compressor as illustrated in Figure 14-7.

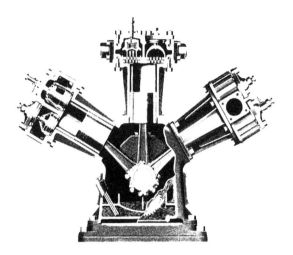

Figure 14–6 Three-piston compressor generates higher vibration levels.

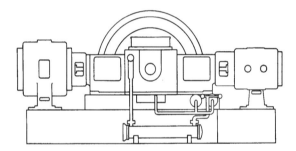

Figure 14–7 Opposed-piston compressor balances piston forces.

Capacity Control

Reciprocating compressors are unique in that they can be used to load-follow. In other words, they can operate continuously when required to meet system demands or can constantly cycle between loaded and unloaded operation without any detrimental effect on the compressor.

Output of reciprocating compressors can be controlled or regulated to match the system demand. Unlike most centrifugal compressors, reciprocating units and most positive-displacement compressors can be operated over a wide range of flow and pressure ranges.

Dynamics

Reciprocating compressors have unique operating dynamics that directly affect their vibration profiles. Unlike centrifugal machinery, but like compressors, reciprocating machines combine rotating and linear motions that generate a complex vibration signature.

Rotational Components

All reciprocating compressors have one or more piston arms attached to one or more crankshafts that provide the motive power to a series of pistons. These crankshafts rotate in the same manner as the shaft in a centrifugal machine. However, their dynamics are somewhat different. The crankshafts will generate all of the normal frequencies of a rotating shaft, i.e., running speed, harmonics of running speed, and bearing frequencies, but the amplitudes will be much higher.

In addition, the relationship of the fundamental (1×) and its harmonics will change. In a normal rotating machine, the fundamental (1×) frequency will normally contain between 60 and 70 percent of the overall or broadband energy generated by the machine train. In reciprocating machines, this profile will change. Two-cycle, reciprocating machines, such as single-acting compressors, will generate a high second harmonic (2×) and harmonics of (2×). Although the fundamental (1×) will be clearly present, it will be at a much lower level.

The shift in vibration profile is the result of the linear motion of the pistons that are used to provide compression of the air or gas. As each piston moves through a complete cycle, it must change direction two times. This reversal of direction generates the higher second harmonic (2×) frequency component.

Reciprocating Components

In a two-cycle machine, all pistons will complete a full cycle each time the crankshaft completes one revolution. Figure 14-8 illustrates the normal action of a two-cycle or single-action compressor. Inlet and discharge valves are located in the clearance space and connected through ports in the cylinder head to the inlet and discharge connections.

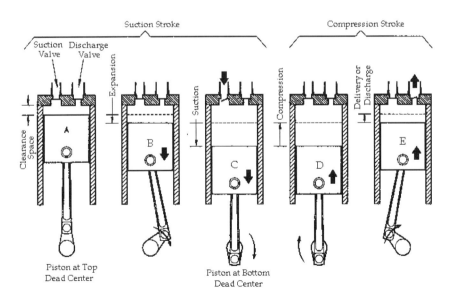

Figure 14–8 Single-acting air compressor cylinder.

During the suction stroke the compressor piston starts its downward stroke and the air under pressure in the clearance space rapidly expands until the pressure falls below that on the opposite side of the inlet valve (Figures 14-8B and 14-8C). This difference in pressure causes the inlet valve to open into the cylinder until the piston reaches the bottom of its stroke (Figure 14-8C).

During the compression stroke the piston starts upward, compression begins, and at point D has reached the same pressure as the compressor intake. The spring-loaded inlet valve then closes. As the piston continues upward, air is compressed until the pressure in the cylinder becomes great enough to open the discharge valve against the pressure of the valve springs and the pressure of the discharge line (Figure 14-8E). From this point to the end of the stroke (Figures 14-8E and 14-8A), the air compressed within the cylinder is discharged at practically constant pressure.

The impact energy generated by each piston as it changes direction will be clearly visible in the vibration profile. Since all pistons complete a full cycle each time the crankshaft completes one full revolution, the total energy of all pistons will be displayed at the fundamental (1×) and second harmonic (2×) locations. In a four-cycle machine, two complete revolutions (720 degrees) are required for all cylinders to complete a full cycle.

Unlike centrifugal shafts, crankshafts have offsets that provide the stroke length for each piston. The orientation of the offsets has a direct effect on the dynamics and vibration amplitudes of the compressor. In an opposed-piston compressor, i.e., where pistons are 180 degrees apart, the impact forces as the pistons change direction are reduced. As one piston reaches top dead center, the opposing piston is also at top dead center. The forces, which are 180 degrees out of phase, tend to cancel or balance the impact forces as the two-piston change direction.

Other configurations, called unbalanced designs, have piston orientations that are neither in phase nor 180 degrees out of phase. In these configurations, the impact forces generated as each piston changes directions are not balanced by an opposite and equal force. As a result, the impact energy and the vibration amplitude are greatly increased.

Failure Modes

Reciprocating compressors, like all reciprocating machines, will normally generate higher levels of vibration than centrifugal machines. In part, this increase is due to the impacts generated as each piston reaches top dead center and bottom dead center of its stroke. The energy levels are also influenced by the unbalanced forces generated by nonopposed pistons and looseness in the piston rods, wrist pins, and journals of the compressor.

In most cases, the dominant frequency will be at the second harmonic (2×) of the main crankshaft's rotating speed. Again, this is the result of the double impact generated by the two changes in direction of each piston during one complete rotation of the crankshaft.

Reciprocating compressors have a history of chronic failures of key components that we will now discuss.

Valves

Valve failure is the predominant failure mode for reciprocating compressors. Because of their high cyclic rate, in excess of 80 million cycles per year, inlet and discharge valves tend to work-harden and crack or break.

Valves can be monitored using traditional vibration monitoring techniques. However, data must be acquired and evaluated using time-domain techniques rather than frequency-domain. Data can be acquired directly from the valve housings using a digital tape recorder, vibration meter, or real-time analyzer. The acquired time trace will provide two indices of valve condition: (1) elapsed time of operation and (2) impacts caused by cracked or broken valve components.

The time required for a valve to completely open and return to the fully closed position should be uniform. The time trace should verify the timing of the valve. If the springs or other valve components have lost memory or are not operating properly, the time interval will change. Comparative analysis of this timing function provides the ability to determine the operating condition of each valve assembly.

Impacts generated by cracked or broken valve components will be clearly visible in the time trace. In normal operation, the valve should generate two impacts for each operating cycle: one when it reaches full open and another when it returns to full close. Both impacts should be of equal amplitude, and the time interval between impacts should be uniform. If components are broken or loose, additional impacts will be observed between these normal impacts. In addition, the impact energy at full open and full closed may increase.

Lubrication System

Poor maintenance of lubricating system components, such as filters and strainers, typically results in premature failure of the lubrication system. Since reciprocating compressors rely on the lubricating system to provide a uniform oil film between closely fitted parts, like those between the piston rings and cylinder wall, partial or complete failure of the lube system will result in catastrophic failure of the compressor.

Although a traditional vibration-monitoring program will detect the metal-to-metal rub that results from lube system failure, it will not prevent damage to the compressor. Therefore, the program should include all components of the lubricating system as well as the compressor. Regular monitoring of the lubricating system components will provide the early warning required to prevent serious compressor damage.

Pulsation

Reciprocating compressors generate pulses of compressed air or gas that is discharged into the piping used to transport it to its points of use. These pulsations can, and often do, generate resonance in the piping system. In addition, these pulses can cause severe damage to other machinery that may be connected to the compressed air system.

Since most compressed air systems do not use pulsation dampers to minimize the pulsations generated by reciprocating compressors, resonance or the impact of these pulses, called *standing waves,* generated by these compressors can result in severe damage to other machine trains included in the compressed air system. Although this is not a failure mode of the compressor, it must be prevented to protect other critical plant equipment. Each time the compressor discharges a volume of compressed air, the air tends to act like a compression spring. It will rapidly expand to fill the available volume of discharge piping, the pulse of high pressure air can result in serious damage.

The full wavelength of the pulsation generated by a double-acting piston design can be obtained by:

$$\lambda = \frac{60a}{2n}$$

or when *a* is 1,135:

$$\lambda = \frac{34,050}{n}$$

where λ = wavelength, feet
        a = speed of sound, feet/second
        n = speed, revolutions/minute

Note: For a single-acting cylinder, the wavelength will be twice as long.

In the example, the compressor running at 1,200 rpm will generate a standing wave of 28.375 feet. In other words, a shock load equivalent to the discharge pressure will be transmitted to any piping or machine connected to the discharge piping that is located within 28 feet of the compressor.

Imbalance

Compressor inertia forces may have two effects on the operating dynamics of reciprocating compressors. The first is a force in the direction of the piston movement. This force will be displayed as impacts as the piston reaches top and bottom dead center of its stroke.

The second is a couple or moment that is developed when there is an offset between the axes of two or more pistons on a common crankshaft. The interrelationship and degree of these forces will depend upon such factors as (1) number of cranks; (2) their longitudinal and angular arrangement; (3) cylinder arrangement; and (4) amount of counterbalancing possible. Two significant vibration periods are set up: the primary at the rotative speed of the compressor and the secondary at twice the rotative speed.

CRANK ARRANGEMENT	FORCES		COUPLES	
	PRIMARY	SECONDARY	PRIMARY	SECONDARY
SINGLE CRANK	F' WITHOUT COUNTERWTS. 0.5 F' WITH COUNTERWTS.	F"	NONE	NONE
TWO CRANKS AT 180° IN LINE CYLINDERS		2F"	F'D W/O COUNTERWTS $\frac{F'D}{2}$ WITH COUNTERWTS	NONE
OPPOSED CYLINDERS	ZERO	ZERO	NIL	NIL
TWO CRANKS AT 90°	1.41 F' WITHOUT COUNTERWTS 0.707 F' WITH COUNTERWTS	ZERO	0.707 F'D W/O COUNTERWTS 0.354 F'D WITH COUNTERWTS	F"D
TWO CYLINDERS ON ONE CRANK CYLINDERS AT 90°	F' WITHOUT COUNTERWTS ZERO WITH COUNTERWTS	1.41 F"	NIL	NIL
TWO CYLINDERS ON ONE CRANK OPPOSED CYLINDERS	2F' WITHOUT COUNTERWTS. F' WITH COUNTERWTS	ZERO	NONE	NIL
THREE CRANKS AT 120°	ZERO	ZERO	3.46F'D W/O COUNTERWTS 1.73 F'D WITH COUNTERWTS	3.48 F"D
FOUR CYLINDERS CRANKS AT 180°	ZERO	4 F"	ZERO	ZERO
CRANKS AT 90°	ZERO	ZERO	1.41 F'D WITHOUT COUNTERWTS 0.707 F'D WITH COUNTERWTS	4.0 F"D
SIX CYLINDERS	ZERO	ZERO	ZERO	ZERO

F' = Primary inertia force in Lbs.
F' = 0.0000284 RN W
F" = Secondary inertia force in Lbs.
F" = R/L x F
R = Crank radius, inches
N = RPM

W = Reciprocating Weight of one cylinder, Lbs.
L = Length of connecting rod, Inches
D = Cylinder center distance

Figure 14–9 Unbalanced inertia forces and couples for various reciprocating compressors.

Although the forces developed are sinusoidal, only the maximum is considered in analysis. Figure 14-9 shows relative values of the inertia forces for various compressor arrangements. The diagrams are plan views with the exception of the fourth arrangement with cylinders at 90° of elevation.

ROTARY COMPRESSORS

The rotary compressor is adaptable to direct drive by induction motors or multicylinder gasoline or diesel engines. The units are compact and relatively inexpensive, and require a minimum of operating attention and maintenance. They occupy a fraction of the space and weight of a reciprocating machine of equivalent capacity. Rotary compressor units are classified into three general groups: sliding-vane type, lobe type, and liquid seal ring type.

Sliding-Vane Compressors

The rotary sliding-vane compressor has as its basic element the cylindrical casing with its heads and rotor assembly. When running at design pressure, the theoretical indicator card is identical to a reciprocating compressor. There is one major difference between a sliding-vane and a reciprocating compressor. The reciprocating unit has spring-loaded valves that open automatically on small pressure differentials between the outside and inside cylinder. The discharge valve, therefore, opens as soon as point 2 (Figure 14-10) is reached and the inlet as point 4 is reached, even though there may be some variation in the discharge pressure.

The sliding-vane compressor has no valves. The times in the cycle when the inlet and discharge open are determined by the location of ports over which the vanes pass. The inlet porting is normally wide and is designed to admit gas up to the point when the pocket between two vanes is the largest. It is closed when the following vane of each pocket passes the edge of the inlet port.

The pocket volume decreases as the rotor turns and the gas is compressed. Compression continues until the leading vane of each pocket uncovers the discharge port. This point must be preset, or built in when the unit is manufactured. Thus, the compressor

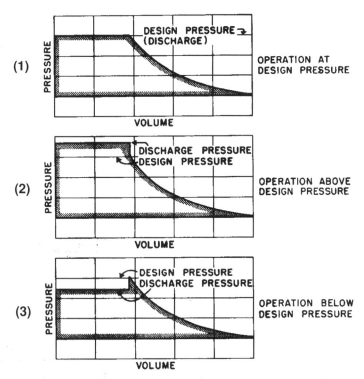

Figure 14–10 Types of theoretical indicator cards obtained by any rotary compressor having built-in porting.

always compresses the gas to *design* pressure, regardless of the pressure in the receiver tank.

The rotary sliding-vane type, as illustrated in Figure 14-11, has longitudinal vanes, sliding radially in a slotted rotor mounted eccentrically in a cylinder. The centrifugal force carries the sliding vanes against the cylindrical case with the vanes forming a number of individual longitudinal cells in the eccentric annulus between the case and rotor. The suction port is located where the longitudinal cells are largest. The size of each cell is reduced by the eccentricity of the rotor as the vanes approach the discharge port, thus compressing the vapor.

Design Fundamentals

The fundamental design considerations of a sliding-vane compressor include the following:

> *Cylinder.* Cast iron is the standard material, but other materials may be used if corrosive conditions exist. The heads contain the bearings and necessary shaft seals. On most standard air compressors, the shaft seals are semimetallic packing in a stuffing box. Commercial mechanical rotary

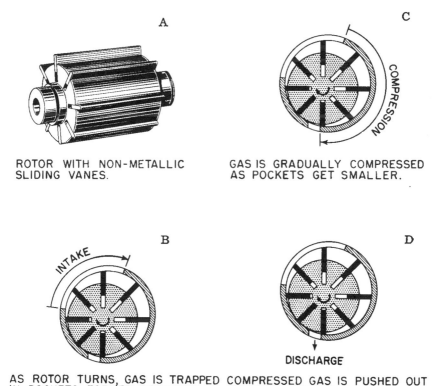

A

ROTOR WITH NON-METALLIC SLIDING VANES.

C

GAS IS GRADUALLY COMPRESSED AS POCKETS GET SMALLER.

B

AS ROTOR TURNS, GAS IS TRAPPED IN POCKETS FORMED BY VANES.

D

DISCHARGE

COMPRESSED GAS IS PUSHED OUT THROUGH DISCHARGE PORT.

Figure 14–11 Rotary sliding-vane air compressor.

seals can be supplied where necessary or desired. Cylindrical roller bearings are standard.

The rotor is usually one piece with the steel shaft using either bar stock or a special forging. Occasionally, the rotor may be a separate iron casting, keyed to the shaft. Vanes are usually asbestos or cotton cloth, impregnated with a phenolic resin. Bronze or aluminum may also be used for vane construction. Each vane fits into a milled slot extending the full length of the rotor and slides radially in and out of this slot once per revolution. Vanes are the most maintenance-prone part in the compressor. There are from 8 to 20 vanes in each rotor, depending on diameter. The greater number of vanes increases compartmentalization and reduces the pressure differential across each vane.

Lubrication. A V-belt-driven force-feed lubricator is used on water-cooled compressors. Oil goes to both bearings and to several points in the cylinder. Ten times as much oil is recommended to lubricate the rotary cylinder as is required for the air cylinder of a corresponding reciprocating compressor. The oil carried over with the gas to the line may be reduced 50 percent with an oil separator on the discharge. Use of an aftercooler ahead of the separator permits removal of 85 to 90 percent of the entrained oil.

Failure Modes

A sliding-vane compressor has failure modes common to vane-type pumps. The dominant vibration profile components will include running speed, vane-pass frequency, and bearing rotational frequencies. In normal operation, all of these frequency components will be low level, with the dominant energy at running speed of the shaft.

Analysis of sliding-vane compressors will use the same techniques and methods as that of any rotating machine train. Load, speed, and process variables must be considered. Common failures of this type of compressor include the following:

Shaft seals. Leakage through the shaft seals should be checked visually once a week or as part of every data acquisition route. Leakage may not extend to the outside of the gland, and if taken off in a vent, the vent discharge should be arranged for easy inspection. Leakage beyond normal is the signal for replacement. Under good conditions, seals have a normal life of 10,000 to 15,000 hours and should be replaced when this service life has been reached.

Vanes. Vanes wear continuously on their outer edges and, to some degree, on the faces that slide in and out of the slots. The vane material is affected somewhat by prolonged heat and gradually deteriorates. Typical life, in 100-psig service, is about 16,000 hours of operation and on low-pressure applications may reach 32,000 hours.

Vane replacement before breakage is extremely important. Breakage while operating can severely damage the compressor and usually forces a complete overhaul and realignment of heads and clearances.

Bearings. In normal service, bearings have a relatively long life, but replacement after about 6 years of operation is recommended. Bearing defects will be displayed in the same manner as any rotating machine train. Inner and outer race defects will be the dominant failure mode, but roller spin may also be present.

Helical Lobe Compressors

The helical lobe, or screw, compressor has two or more mating sets of lobe-type rotors mounted in a common housing. The male lobe or rotor is usually direct-driven by an electric motor. The female or mating rotor is driven by a helical gear set that is mounted on the outboard end of the rotor shafts. The gears provide both motive power for the female rotor and absolute timing between the rotors.

The rotor set has extremely close mating clearance, about 0.5 mils, but without metal-to-metal contact (Figure 14-12). Most of the rotary compressors are designed for *oil-free* operation. In other words, no oil is used to lubricate or seal the rotors. Instead, lubrication is limited to the timing gears and bearings that are outside the air chamber. As a result, maintaining proper clearance between the two rotors is critical.

This type of compressor is classified as a *constant-volume, variable-pressure* machine and is quite similar to the vane-type rotary in general characteristics. Both have a built-in compression ratio.

Helical-lobe compressors are best suited for base-load applications where they can provide a constant volume and pressure of discharge gas. The only recommended

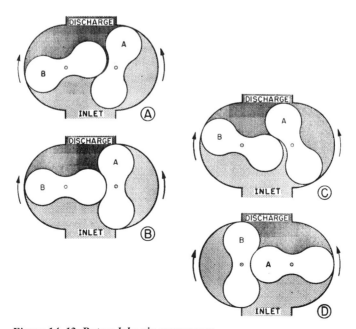

Figure 14–12 Rotary lobe air compressor.

method of volume control is the use of variable-speed motors. With variable-speed drives, capacity variations can be obtained with a proportionate reduction in speed. A 50 percent speed reduction is the maximum permissible control range.

Helical-lobe compressors are not designed for frequent or constant cycles between load and no-load operation. Each time the compressor unloads, the rotors tend to thrust axially. Even though the rotors have a substantial thrust bearing and in some cases a balancing piston, to counteract axial thrust, each time the compressor unloads, the axial clearance is increased. With time, this clearance will increase enough to permit a dramatic increase in the impact energy created by axial thrust during the transient from loaded to unloaded conditions. In extreme cases, the energy will be enough to physically push the rotor assembly through the compressor housing.

Design Considerations

These units are primarily designed to operate dry, with nonlubricated compression chambers sealed at leakage points by close clearances. Since these clearances must be kept open at all times, any operational factors affecting them must be limited. If not, there might be sufficient expansion or distortion to cause rotor contact and consequent damage. These and other limiting factors include the following:

> *Discharge temperature.* Discharge temperatures must be limited to prevent excessive casting distortion and rotor growth. If either occurs, the clearance between rotating parts will decrease and metal-to-metal contact will take place. Since the rotors typically rotate at speeds between 3,600 and 10,000 rpm, metal-to-metal contact normally results in instantaneous, catastrophic failure of the compressor.
>
> *Temperature rise.* A limit is sometimes set to prevent excessive relative distortion between the inlet and discharge ends of the casing and excessive rotor growth. The higher-pressure units are water-jacketed for uniform casing temperature. Rotors may also be cooled to permit a higher operating limit.
>
> *Compression ratio.* Compression ratio and maximum inlet temperature determine the maximum discharge temperature of rotary screw compressors.
>
> *Pressure rise.* Pressure rise is the differential pressure across the compressor, i.e., the difference between discharge and inlet pressure. This acts to deflect the rotor and shaft. Any bending of the rotor or its shaft changes the original clearance.
>
> *Bearing loads.* Changes in differential pressures, caused by variations in either inlet or discharge conditions, will result in a change in the bearing load zone. If there is a change in either the inlet or discharge conditions, i.e., temperature, volume, or pressure, the rotors tend to become unstable. As a result

of this instability, the load zones are changed in the shaft support bearings. The normal result is premature wear and/or failure of the bearings.

Machine Dynamics

Rotary screw compressors are designed to deliver a constant volume and pressure of air or gas. Any variation in either the inlet or discharge conditions will result in rotor instability that can result in instantaneous failure of the compressor. In normal operation, the compressor will generate a vibration profile that will include the following:

> *Rotor speeds.* In a single-stage rotary screw compressor, there will be two mating rotors. The male will be directly driven by an electric motor. Typically, this input speed will be between 1,800 and 3,600 rpm. The mating female rotor will normally operate at a higher speed than the male. The normal speed range for the female rotor is between 3,000 and 5,000 rpm.

> In two-stage rotary screw compressors, there is a second set of mated rotors that extract compressed air from the first set and further compress the gas. Speeds on these rotors can be as much as 10,000 rpm.

> Each of these rotors will generate a fundamental (1×) frequency component at their actual running speed. Speed should be relatively uniform, but slight variations resulting from load changes may be observed.

> *Rotor mesh.* The passing frequency generated by the meshing of the male and female rotors will generate a vibration frequency in the same manner as a gear set or fan blades. In a normal application, the energy generated by this passing frequency should be very low, but will increase dramatically if any process instability occurs.

> *Gear mesh.* The helical timing gears that are used to synchronize the male and female rotors will generate a gear mesh profile. The rules used to monitor and evaluate a standard single-reduction helical gear set should be used for this application.

> *Bearing frequencies.* Most rotary screw compressors use rolling-element bearings on both rotor shafts. The normal configuration has a large thrust or fixed bearing located on the outboard end of each shaft and float bearing on the inboard ends. These bearings should be monitored in the same manner as any rolling-element bearing.

Failure Modes

The most common reason for compressor failure or component damage is process instability. These units are extremely susceptible to any change in either inlet or discharge conditions. A slight variation in pressure, temperature, or volume can result in instantaneous failure.

The following indices of instability and potential problems should be used:

> *Rotor mesh.* In normal operation, the energy generated by the meshing of the male and female rotors will be very low level. As in the case of vane or blade passing, the peak should be narrow and have low amplitude.
>
> If the process becomes unstable, the energy surrounding the rotor meshing frequency will increase. Both the amplitude of the meshing frequency and the width of the peak will increase. In addition, the noise floor surrounding the meshing frequency will become more active. This *white noise* will be similar to that observed in a cavitating pump or unstable fan.
>
> *Axial movement.* The normal tendency of both the rotors and the helical timing gears is to generate axial movement or thrusting of the shafts. The extremely tight clearances between the male and female rotors will not tolerate any excessive axial shaft movement. Therefore, a primary monitoring parameter must be the amount of axial movement that is present in the compressor.
>
> Axial measurements should be acquired and monitored on both rotor assemblies. If there is any increase in the vibration amplitude in these measurement points, there is a high probability of compressor failure.
>
> *Thrust bearings.* Although process instability can affect both the fixed and float bearings, the thrust bearing is more likely to show early degradation as a result of process instability or abnormal compressor dynamics. These bearings should be monitored closely. Any degradation or suggestion of excessive axial clearance should be corrected immediately.
>
> *Gear mesh.* The gear mesh profile will also provide an indication of prolonged instability in the compressor. Deflection of the rotor shafts will change the wear pattern on the helical gear sets. This change in pattern will increase the backlash in the gear mesh and result in higher vibration levels as well as an increase in thrusting.

Liquid Seal Ring Compressors

The rotary liquid seal ring type, as illustrated in Figure 14-13, features a forward-inclined, open impeller in an oblong cavity filled with liquid. As the impeller rotates, the centrifugal force causes the seal liquid to collect at the outer edge of the oblong cavity. Because of the oblong configuration of the compressor case, large longitudinal cells are created and reduced to smaller ones. The suction port is positioned where the longitudinal cells are the largest, and the discharge port where they are smallest, thus causing the vapor within the cell to compress as the rotor rotates. The rotary liquid seal compressor is frequently used in specialized applications for the compression of extremely corrosive and exothermic gases and is commonly used in commercial nuclear plants as a means of establishing initial condenser vacuum.

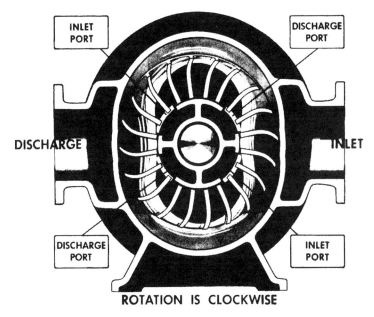

INLET
PORT

DISCHARGE
PORT

DISCHARGE

INLET

DISCHARGE
PORT

INLET
PORT

ROTATION IS CLOCKWISE

Figure 14–13 Rotary liquid seal ring air compressor.

Design Considerations

The rotary liquid-piston or liquid-ring compressor uses a rotor with multiple forward turned blades turning about a central cone containing inlet and discharge ports. The blades drive a captive ring of liquid around the inside of an elliptical casing. The basic elements are the casing heads and rotor assembly.

A certain amount of liquid is trapped between adjacent blades, and as the rotor turns, the liquid face moves in and out of this space because of the casing shape. This creates a liquid piston. Porting in the central cone is built-in and fixed. There are no valves.

There is compression within the pockets or chambers between the blades before the discharge port is uncovered. Since the port location must be designed and built for a specific compression ratio, it will tend to operate above or below the design pressure.

The cooling of liquid-ring compressors is direct rather than through the walls of the casing. The required additional cooling liquid is fed into the casing where it comes into direct contact with the gas being compressed. The excess liquid is discharged with the gas. The discharged mixture is passed through a conventional baffle or centrifugal type separator to remove the free liquid. Because of the intimate contact of gas and liquid, the final discharge temperature can be held close to the temperature of the inlet cooling water. However, the discharge gas is saturated at the discharge temperature of the compressed liquid.

The amount of liquid that may be passed through the compressor is not critical and can be varied to obtain the desired results. The unit will not be damaged if a large quantity of liquid inadvertently enters its suction. Lubrication is required only in the bearings, which are generally located external to the casing. The liquid itself acts as a lubricant, sealing medium, and coolant for the stuffing boxes.

Dynamics and Failure Modes

The liquid-ring compressor operates much like any rotation machine train. It will generate the same dynamics and vibration profile as a centrifugal pump or fan. Although the compressor is positive displacement, i.e., discharges the same volume for each rotation, its operating dynamics are centrifugal.

DYNAMIC (CENTRIFUGAL) COMPRESSORS

Compression in any dynamic or centrifugal compressor depends on the transfer of energy from a rotating set of blades to a gas. The rotor accomplishes this energy transfer by changing the momentum and pressure of the gas. The momentum, i.e., kinetic energy, then is converted into useful pressure energy by slowing the gas down in a stationary diffuser or another set of blades. Dynamic compressors are generally designated as centrifugal, axial, or mixed-flow. Although these compressors are constructed differently, the same basic aerodynamic design theory applies to all three.

Centrifugal Compressors

The *centrifugal* designation is used when the gas flow is radial and the energy transfer is predominantly due to a change in the centrifugal forces acting on the gas.

The centrifugal compressor, originally built to handle only large volumes of low-pressure gas and air (maximum of 40 psig), has been developed to enable it to move large volumes of gas with discharge pressures up to 3,500 psig. However, centrifugal compressors are now most frequently used for medium-volume and medium-pressure air delivery. One advantage of a centrifugal pump is the smooth discharge of the compressed air.

The centrifugal force utilized by the centrifugal pump is the same force utilized by the centrifugal compressors. The air particles enter the eye of the impeller, designated D in Figure 14-14. As the impeller rotates, air is thrown against the casing of the compressor. The air becomes compressed as more and more air is thrown out to the casing by the impeller blades. The air is pushed along the path designated A, B, and C in Figure 14-14. The pressure of the air is increased as it is pushed along this path.

There may be several stages to a centrifugal air compressor, as in the centrifugal pump, and the result would be the same; a higher pressure would be produced. The air compressor is used to create compressed or high-pressure air for a variety of uses. Some of its uses are pneumatic control devices, pneumatic sensors, pneumatic valve operators, pneumatic motors, and starting air for diesel engines.

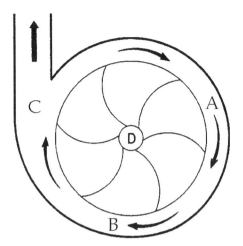

Figure 14–14 Simplified centrifugal compressor.

Design Considerations

The centrifugal compressor has a limited stable operating range. This may affect the economics of operation at partial load. The minimum capacity may vary from 45 to 90 percent of rated capacity. The centrifugal must be selected for the worst combination of conditions that exist at a given time and must be controlled to meet other requirements.

Operating speeds are high compared to other compressors. Speeds between 50,000 and 100,000 rpm are common in some compressor designs. For example, the fourth-stage impeller in a typical bullgear-type compressor, such as the Ingersoll-Rand Centac (Figure 14-15), will normally operate above 60,000 rpm.

These compressors are well suited for direct connection to steam or gas turbine drivers, which permit variable speed control. Variable speed control is the only recommended means of volume or load control on centrifugal compressors. They are designed to be base-loaded and to operate continuously at constant volume and pressure.

In normal operating conditions, i.e., base-loaded, these machines have a high availability factor. They frequently operate without shutdown for 2 to 3 years.

Dynamics and Failure Modes

The dynamics of a centrifugal compressor are the same as for other centrifugal machine trains. The dominant forces and vibration profiles will be identical to those of a pump or fan. However, the effects of variable load and other process variables, such as temperature and inlet/discharge pressure, are more pronounced than in other rotating machines.

Monitoring and diagnostic techniques and methods should be the same as for a similar pump, gearbox, or other similar rotating machine train. Radial measurements

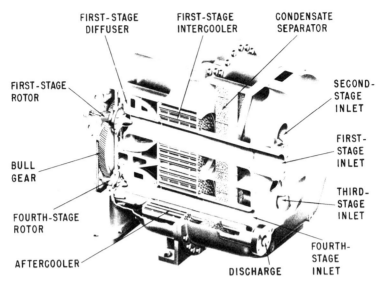

FIRST-STAGE DIFFUSER

FIRST-STAGE INTERCOOLER

CONDENSATE SEPARATOR

FIRST-STAGE ROTOR

SECOND-STAGE INLET

FIRST-STAGE INLET

BULL GEAR

THIRD-STAGE INLET

FOURTH-STAGE ROTOR

FOURTH-STAGE INLET

AFTERCOOLER

DISCHARGE

Figure 14–15 Bullgear-type compressor.

should be located in the planes that will provide the earliest indication of shaft displacement created by abnormal operation.

Aerodynamic instability is the most common failure mode of centrifugal compressors. Variable demand or restrictions of the inlet airflow are the more common sources of this instability. Even slight variations in either of these process variables can result in dramatic changes in the operating stability of the compressor.

Entrained liquid and solids can also affect operating life of centrifugal compressors. Controlled liquid injection for cooling and cleaning may be considered in design. When dirty air must be handled, open-type impellers must be used. Their open design will permit the capability to handle a moderate amount of dirty or other solids in the inlet air supply. However, inlet filters are recommended for all applications.

The actual dynamics of centrifugal compressors is driven by their design. A variety of designs ranging from overhung rotors to bullgears are commonly used:

> *Overhung or cantilever.* The cantilever design is more susceptible to process instability than centerline compressors. Figure 14-16 illustrates a typical cantilever design.
>
> The overhung design of the rotor, i.e., no outboard bearing, increases the potential for radical shaft deflection. Any variation in laminar flow of inlet or discharge air will force the shaft to bend or deflect from its true centerline. As a result, mode shape of the shaft must be monitored closely. If there is any indication of deflection, the air control system, i.e., dampers or

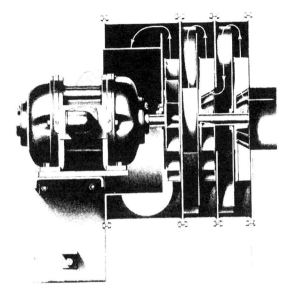

Figure 14–16 Cantilever design is susceptible to instability.

flow-control valves, must be inspected and the system balanced to permit stable operation.

Centerline design. Centerline designs, such as the horizontal and vertical split-case, are more stable over a wider operating range, but should not be operated in a variable-demand system. Figure 14-17 illustrates the normal airflow pattern through a horizontal split-case compressor. Inlet air enters the first stage of the compressor where the pressure and velocity are increased. The partially compressed air is routed to the second stage where the velocity is further increased. This process can be continued, by adding additional stages, until the final discharge pressure is achieved.

Two factors are critical to the normal operation of these compressors: the fact that laminar flow must be maintained through all stages, and the design of the impeller.

There are two impeller designs used in centrifugal compressors: inline and back-to-back. In the former, all impellers face in the same direction. In the latter, impeller direction is reversed in adjacent stages.

Inline impellers—compressors with all impellers facing in the same direction—will, by design, generate substantial axial forces. The axial pressure generated by each impeller will be added for all stages, and as a result massive axial loads will be transmitted to the fixed bearing. Thus, most of these compressors will use either a Kingsbury thrust bearing or a balancing piston to resist this axial thrusting. Figure 14-18 illustrates a typical *balancing piston.*

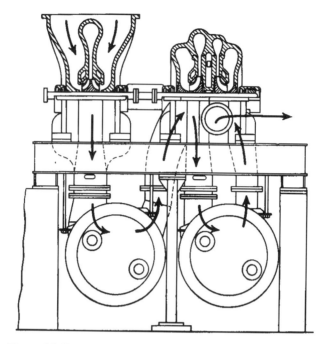

Figure 14–17 Air flow through a centerline compressor.

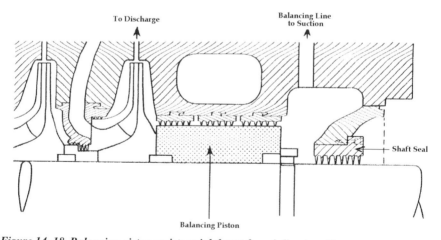

Figure 14–18 Balancing piston resists axial thrust from inline impellers.

All compressors that use inline impellers must be monitored closely for axial thrusting. If the compressor is subjected to frequent or constant unloading, the axial clearance will be increased. Ultimately, this will lead to catastrophic failure of the compressor.

Opposed impellers are a different matter. Reversing the direction of alternating impellers means that the axial forces generated by each impeller can be minimized by opposing force of the adjacent impeller. In effect, the opposed impellers tend to cancel the axial forces generated by the preceding stage.

This design is more stable and should not generate measurable axial thrusting. Therefore, these units contain a normal float and fixed rolling-element bearing.

Axial Compressors

The *axial* designation is used when the gas flow is parallel to the compressor shaft. Energy transfer is caused by the action of a number of rows of blades or a rotor, each row followed by a fixed row fastened to the casing.

The axial-flow dynamic compressor is shown in Figure 14-19. It is essentially a large-capacity, high-speed machine with characteristics quite different from those of the centrifugal compressor. Each stage consists of two rows of blades, one row rotating and the next row stationary. The rotor blades impart velocity and pressure to the gas as the rotor turns, the velocity being converted to pressure in the stationary blades. Frequently about half the pressure rise is generated in the rotor blades and half in the stator. Gas flow is predominantly in an axial direction and there is no appreciable vortex action.

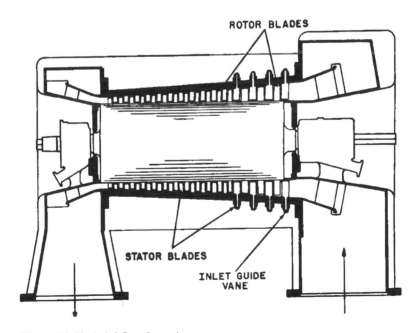

Figure 14–19 Axial-flow dynamic compressor.

Dynamics

The dynamic forces generated by an axial-flow compressor are also similar to those of other rotating machinery. The major difference is that the dominant orientation of process forces is parallel to the shaft. As a result, the axial forces and vibration profile will be dominant.

Special attention must be given to the axial movement and vibration of axial-flow compressors. A variation in the operating dynamics and stability of this design will be dominant in the axial direction. Radial readings will help in understanding the operating conditions, but will be less definitive than in other designs.

Mixed-Flow Compressors

The *mixed-flow* designation is used when the gas flow is between radial and axial. It combines the design features of each with characteristics also lying between the two. This type is not applied as frequently as the others. Because of the long length required for each stage, this type is generally not found in multistage designs.

COMPRESSOR SELECTION

Air power compressors generally operate at pressures of 500 psig or lower, with the majority in the range of 125 psig or less. All major types of compressors (i.e., reciprocating, vane, helical lobe, and dynamic) are used for this type of service. Choice is limited somewhat by capacity at 100 psig of about 10,500 cubic feet per minute but can be built to approximately 28,000 cfm. The vane-type rotary has an upper listed size of 3,700 cfm as a twin unit and the helical lobe rotary can be used to nearly 20,000 cfm. The centrifugal can be built to very large sizes. It is currently offered in the proven, moderate speed designs starting at a minimum of about 5,000 cfm.

Selection Criteria

The following guidelines should be used for the selection process. Although the criteria listed are not all-inclusive, they will provide definition of the major considerations that should be used to select the best compressor for a specific application.

Application

The mode of operation of a specific application should be the first consideration. The inherent design of each type of compressor defines the acceptable operating envelope or mode of operation that it can perform with reasonable reliability and life cycle costs. For example, a bullgear-type centrifugal compressor is not suitable for load-following applications but will prove exceptional service in constant-load and -volume applications.

Load factor is the ratio of actual compressed air output, while the compressor is operating, to the rated full-load output during the same period. It should never be 100 percent, a good rule being to select an installation for from 50 to 80 percent load factor, depending on the size, type, and number of compressors involved. Proper use of load factor results in: more uniform pressure, a cooling-off period, less maintenance, and ability to increase use of air without additional compressors.

Load factor is particularly important with air-cooled machines where sustained full-load operation results in an early buildup of deposits on valves and other parts. This buildup increases the frequency of maintenance required to maintain compressor reliability. Intermittent operation is always recommended for these units. The frequency and duration of *unloaded* operation depend on the type, size, and operating pressure of the compressor. Air-cooled compressors for application at pressures higher than 200 psig are usually rated by a rule that states that the *compressing time* shall not exceed 30 minutes or less than 10 minutes. Shutdown or unloaded time should be at least equal to compression time or 50 percent.

Rotary screw compressors are exceptions to this 50 percent rule. Each time a rotary screw compressor unloads, both the male and female rotors instantaneously shift axially. These units are equipped with a balance piston or heavy-duty thrust bearing that is designed to absorb the tremendous axial forces that result from this instantaneous movement, but they are not able to fully protect the compressor or its components. The compressor's design accepted the impact loading that results from this unload shifting and incorporated enough axial strength to absorb a normal unloading cycle. If this type of compressor is subjected to constant or frequent unloading, as in a load-following application, the cycle frequency is substantially increased and the useful life of the compressor is proportionally reduced. There have been documented cases where either the male or female rotor actually broke through the compressor's casing as a direct result of this failure mode.

The only compressor that is ideally suited for load-following applications is the reciprocating type. These units have an absolute ability to absorb the variations in pressure and demand without any impact on either reliability or life-cycle cost. The major negative of the reciprocating compressor is the pulsing or constant variation in pressure that is produced by the reciprocating compression cycle. Properly sized accumulators and receiver tanks will resolve most of the pulsing.

Life-Cycle Costs

All capital equipment decisions should be based on the true or life-cycle cost of the system. Life-cycle cost includes all costs that will be incurred, beginning with specification development before procurement to final decommissioning cost at the end of the compressor's useful life. In many cases, consideration is given only to the actual procurement and installation cost of the compressor. Although these costs are important, they represent less than 20 percent of the life-cycle cost of the compressor.

The cost evaluation must include the recurring costs, such as power consumption and maintenance, that are an integral part of day-to-day operation. Other costs that should be considered include training of operators and maintenance personnel who must maintain the compressor.

15

AIR DRYERS

Air entering the first stage of any air compressor carries with it a certain amount of native moisture. This is unavoidable, although the quantity carried will vary widely with the ambient temperature and relative humidity. Figure 15-1 shows the effect of ambient temperature and relative humidity on the quantity of moisture in atmospheric air entering a compressor at 14.7 psia. Under any given conditions, the amount of water vapor entering the compressor per 1,000 cubic feet of mixture may be approximated from these curves.

In any air–vapor mixture, each component has its own partial pressure and the air and the vapor are each indifferent to the existence of the other. It follows that the conditions of either component may be studied without reference to the other. In a certain volume of mixture, each component fills the full volume at its own partial pressure. The water vapor may saturate this space, or it may be superheated.

As this vapor is compressed, its volume is reduced; at the same time the temperature automatically increases. As a result, the vapor becomes superheated. More pounds of vapor are now contained in one cubic foot than when the vapor originally entered the compressor.

Under the laws of vapor, the maximum quantity of a particular vapor a given space can contain is dependent solely upon the vapor temperature. As the compressed water vapor is cooled, it will eventually reach the temperature at which the space becomes saturated, now containing the maximum it can hold. Any further cooling will force part of the vapor to condense into its liquid form—water.

The curves contained in Figure 15-2 show what happens over a wide range of pressures and temperatures. However, these are saturated vapor curves based on starting with 1,000 cubic feet of *saturated* air. If the air is not saturated at the compressor's inlet, and it usually is not, use Figure 15-3 to obtain the initial water vapor weight

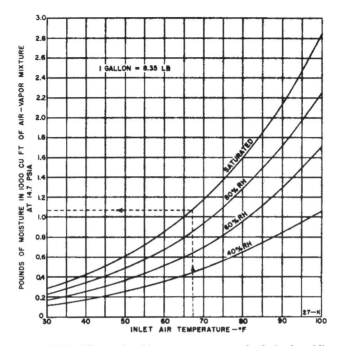

Figure 15–1 Effects of ambient temperature and relative humidity.

entering the system per 1,000 cubic feet of compressed air. By reading left on Figure 15-4 from the juncture of the final pressure and final temperature, obtain the maximum weight of vapor that this same 1,000 cubic feet can hold after compression and cooling to saturation. If the latter is less than the former, the difference will be condensed. If the latter is higher, there will be no condensation. It is clearly evident that the lower the temperature and the greater the pressure of compressed air, the greater will be the amount of vapor condensed.

PROBLEMS CAUSED BY WATER IN COMPRESSED AIR

Few plant operators need to be told of the problems caused by water in compressed air. They are most apparent to those who operate pneumatic tools, rock drills, automatic pneumatic powered machinery, paint and other sprays, sandblasting equipment, and pneumatic controls. However, almost all applications, particularly of 100-psig power, could benefit from the elimination of water carryover. The principal problems might be summarized as follows:

1. Washing away of required lubrication
2. Increase in wear and maintenance
3. Sluggish and inconsistent operation of automatic valves and cylinders
4. Malfunctioning and high maintenance of control instruments

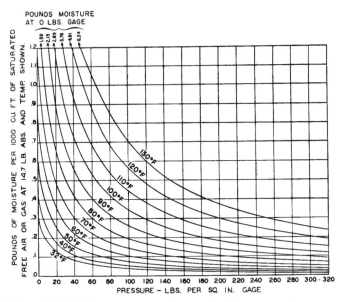

Figure 15–2 Moisture remaining in saturated air or gas when compressed isothermally to pressure shown.

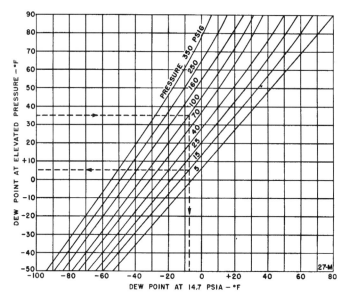

Figure 15–3 Dewpoint conversion chart.

5. Spoilage of product by spotting in paint and other types of spraying
6. Rusting of parts that have been sandblasted
7. Freezing in exposed lines during cold weather
8. Further condensation and possible freezing of moisture in the exhaust of those more efficient tools that expand the air considerably

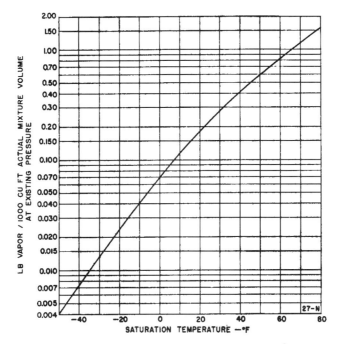

Figure 15–4 Moisture in air at any pressure for 100 ft³ of actual volume at existing pressures.

A fact to remember is that water vapor, *as vapor*, does no harm in most pneumatic systems. It is only when the vapor condenses and remains in the system as a liquid that problems exist. The goal, therefore, is to condense and remove as much of the vapor as is economically possible.

In conventional compressed air systems, vapor and liquid removal is limited. Most two-stage compressors will include an intercooler between stages. On air-cooled units for 100 to 200 psig service, the air between stages is not cooled sufficiently to cause substantial liquid drop out and provision is not usually made for its removal. Water-cooled intercoolers used on larger compressors will usually cool sufficiently to condense considerable moisture at cooler pressure. Drainage facilities must always be provided and used. Automatic drain traps are normally included to drain condensed water vapor.

All compressed-air systems should always include a water-cooled aftercooler between the compressor and receiver tank. In normal summer conditions, properly designed and maintained aftercoolers at 100 psig will condense up to 70 percent of the vapor entering the system. Most of this condensation will collect in the aftercooler or the receiver tank. Therefore, both must be constantly drained.

The problem with a conventional system that relies on heat exchangers (i.e., aftercoolers) for moisture removal is temperature. The aftercooler will only remove liquids that have condensed at a temperature between the compressed air and cooling water temperature. In most cases, this differential will be about 20 to 50 degrees lower than the compressed air temperature, or around 70 to 90°F. As long as the compressed air remains at or above this temperature range, any vapor that it still contains will stay in a vapor or gaseous state. However, when the air temperature drops below this range, additional vapor will condense into water.

DRIED AIR SYSTEMS

This system involves processing the compressed air or gas after the aftercooler and receiver to further reduce moisture content. This requires special equipment, a higher first cost, and a higher operating cost. These costs must be balanced against the gains obtained. They may show up as less wear and maintenance of tools and air-operated devices, greater reliability of devices and controls, and greater production as a result of fewer outages for repairs. In many cases, reduction or elimination of product spoilage or a better product quality may also result.

The degree of drying desired will vary with the pneumatic equipment and application involved. The aim is to eliminate further condensation in the air lines and pneumatic tools or devices. Prevailing atmospheric conditions also have an influence on the approach that is most effective. In many 100-psig installations, a dew point at line pressure of 500 to 350°F is adequate. In other applications, such as instrument air systems dew points of –500°F is required.

Terminology involves drier outlet dew point at the *line* pressure or the pneumatic circuit. This is the saturation temperature of the remaining moisture contained in the compressed air or gas. If the compressed gas temperature is never reduced below the outlet dew point beyond the drying equipment, there will be no further condensation.

Another value sometimes involved when the gas pressure is reduced before it is used is the dew point at that lower pressure condition. A major example is the use of 100 psig (or higher) gas reduced to 15 psig for use in pneumatic instruments and controls. This dew point will be lower because the volume involved increases as the pressure is decreased. The dew point at atmospheric pressure is often used as a reference point for measurement of drying efficiency. This is of little interest when handling *compressed* air or gas.

Figure 15-5 enables one to determine dew point at reduced pressure. The left scale shows the dew point at the elevated pressure. Drop from the intersection of this value and the elevated pressure line to the reduced pressure line and then back to the left to read the dew point at the reduced pressure.

Figure 15-6 shows graphically the amount of moisture remaining in the vapor form when the air–vapor mixture is conditioned to a certain dew point. This curve is based on a volume of 1,000 cubic feet or an air vapor mixture *at its total pressure*. For example, 1,000 cubic feet at 100 psig air at 500°F and 1,000 cubic feet of 15 psig air at 50°F will hold the same vapor at the dew point. However, 1,000 cubic feet at 100

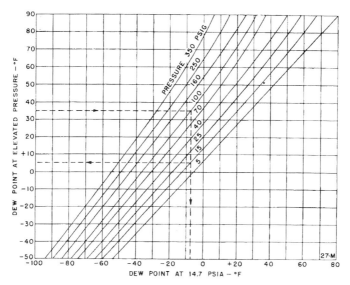

Figure 15–5 Dewpoint conversion chart.

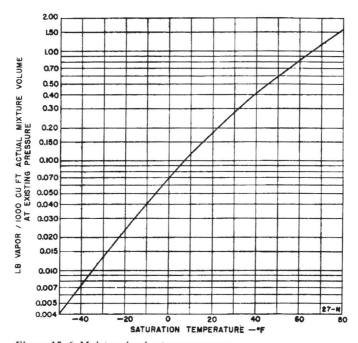

Figure 15–6 Moisture in air at any pressure.

psig and 50°F reduced to 15 psig will become 3,860 cubic feet at 50°F. As a result, it is now capable of holding 3.86 times as much vapor and the dew point will not be reached until the mixture temperature is lowered to its saturation temperature.

DRYING METHODS

There are three general methods of drying compressed air: chemical, adsorption, and refrigeration. In all cases, aftercooling and adequate condensation removal must be done ahead of this drying equipment. The initial and operating costs and the results obtained vary considerably.

These methods are primarily for water vapor removal. Removal of lubricating oil is secondary, although all drying systems will reduce its carryover. It should be understood that complete elimination of lubricating oil, particularly in the vapor form, is very difficult and that when absolutely oil-free air is required, some form of nonlubricated compressor is the best guaranteed method.

Chemical Dryers

Chemical dryers are materials that combine with or absorb moisture from air when brought into close contact. There are two general types. One, using deliquescent material in the form of pellets or beads, is reputed to obtain a dew point, with 700°F inlet air to the dryer, of between 35 and 50°F, depending on the specific type of deliquescent material. The material turns into a liquid as the water vapor is absorbed. This liquid must be drained off and the pellets or beads replaced periodically. Entering air above 900°F is not generally recommended.

The second type of chemical dryer utilizes an ethylene glycol liquid to absorb the moisture. Standard dew point reduction claimed is 400°F, but greater reductions are said to be possible with special equipment. The glycol is regenerated (i.e., dried) in a still using fuel gas or steam as a heating agent. The released moisture is vented to atmosphere. The regenerated glycol is recirculated by a pump through a water-cooled heat exchanger that lowers the glycol temperature before returning to the dryer vessel.

Adsorption

Adsorption is the property of certain extremely porous materials to hold vapors in the pores until the desiccant is either heated or exposed to a drier gas. The material is a solid at all times and operates alternately through drying and reactivation cycles with no change in composition. Adsorbing materials in principal use are activated alumina and silica gel. Molecular sieves are also used. Atmospheric dew points of −1,000°F are readily obtained using adsorption.

Reactivation or regeneration is usually obtained by diverting a portion of the already dried air through a reducing valve or orifice, reducing its pressure to atmospheric, and passing it through the wet desiccant bed. This air, with the moisture it has picked up from the saturated desiccant bed, is vented to atmosphere. The diverted air may vary from 7 to 17 percent of the main stream flow, depending upon the final dew point desired from the dryer. Heating the activating air prior to its passing through the desiccant bed, or heating the bed itself, is often done to improve the efficiency of the regeneration process. This requires less diverted air since each cubic foot of diverted air will carry much more moisture out of the system. Other modifications are also available to reduce or even eliminate the diverted air quantity.

Refrigeration

The use of refrigeration for drying compressed air is growing rapidly. It has been applied widely to small installations, to sections of larger plants, and even to entire manufacturing plant systems. Refrigerated air dryers have been applied to the air system both before and after compression. In the *before compression* system, the air must be cooled to a lower temperature for a given *final line pressure dew point*. This takes more refrigeration power for the same end result. Partially offsetting this is a saving in air compressor power per 1,000 cubic feet per minute (cfm) of *atmospheric* air compressed due to the reduction in volume at the compressor inlet caused by the cooling and the removal of moisture. There is also a reduction in discharge temperature on single-stage compressors that may at times have some value. An atmospheric (inlet) dew point of 350°F is claimed.

When air is refrigerated *following compression*, two systems have been used. Flow of air through directly refrigerated coils is used predominately in the smaller and moderate-sized systems. These are generally standardized for cooling to 350°F, which is the dew point obtained at line pressure. Figure 15-7 diagrams the equipment furnished in a small self-contained direct-refrigeration dryer.

The larger systems chill water that is circulated through coils to cool the air. A dew point at line pressure of about 500°F is obtainable with this method. Figure 15-8 illustrates a typical system of a *chiller-dryer* unit. The designs shown are regenerative since the incoming air is partially cooled by the outgoing air stream. This reduces the size and first cost of the refrigeration compressor and exchanger. It also reduces

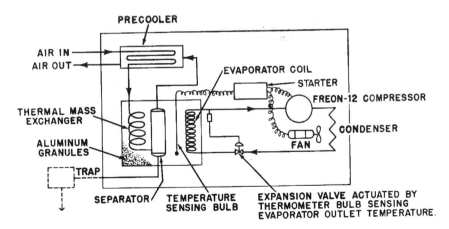

Figure 15–7 Diagram of equipment furnished in small self-contained direct-refrigeration dryer.

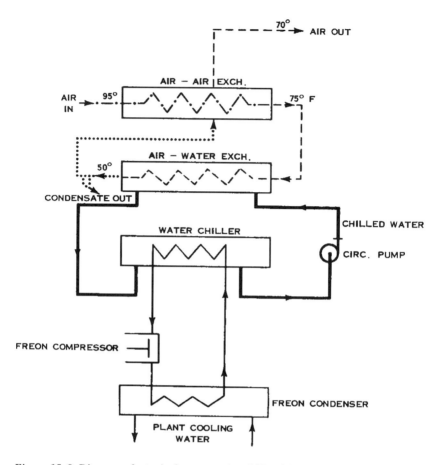

Figure 15–8 Diagram of a typical regenerative chiller-drier.

power cost and reheats the air returning to the line. Reheating of the air after it is dried has several advantages: the air volume is increased and less *free air* is required; chance of line condensation is still further reduced; and sweating of the cold pipe leaving the dryer is eliminated. Reheating dryers seldom need further reheating.

Combination Systems

The use of a combination dryer should be investigated when a very low dew point is necessary. Placing a refrigeration system ahead of an adsorption dryer will let the more economical refrigeration unit remove most of the vapor and reduce the load on the desiccant.

DRY AIR DISTRIBUTION SYSTEM

Most plants are highly dependent on their compressed air supply, and it should be ensured that the air is in at least reasonable condition at all times, even if the drying sys-

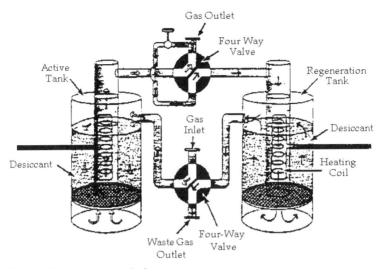

Figure 15–9 Desiccant air dryer.

tem is out of use for maintenance or repair. It is possible that the line condensation would be so bad that some air applications would be handicapped or even shut down if there were no protection. A vital part of all the effort to separate water in the conventional compressed air system is also the trapping of dirt, pipe scale, and other contaminants. This is still necessary with a dried air system. As a minimum, all branch lines should be taken off the top of the main and all feeder lines off the top of branch lines.

Absolute prevention of line freezing can be obtained only when the dew point of the line air is below any temperature to which it may be exposed. Freezing is always possible if there is line condensation. For example, when air lines are run outdoors in winter weather or pass through cold storage rooms, the ambient temperatures will change the dew point and cause any moisture in the air to condense and freeze.

The dryer unit has an air inlet, an air outlet, a waste air outlet, two heater coils, and two four-way reversing valves (see Figure 15-9). Although the illustration shows the two tanks in a functioning mode, they are universal in operation. By this, we mean that this changes the tanks over when the active tank becomes totally saturated, and the tank that was on the regeneration cycle then becomes the active unit.

With this process, there will always be one tank in active service, and one on regeneration or stand-by. A timer governs the changeover process. The timer changes the position of the four-way, or reversing, valves. This operation reverses the airflow through the tanks; the tank that was regenerated will now dry the air, while the saturated tank will be regenerated.

16

AIR RESERVOIR (RECEIVER)

Air receivers, being simple volume tanks, are not often thought of as highly engineered items, but the use of simple engineering with receivers can reduce equipment costs. A pertinent example not uncommon in industry is the intermittent requirement for a fairly large volume of air at moderate pressure for a short period of time. Some boiler soot-blowing systems are in this class. The analysis necessary to arrive at the most economical equipment often involves the storage of air at high pressure to supplement the compressor's output when the demand requires. The following example is somewhat extreme, but emphasizes the need for proper receiver selection.

SIZING A RECEIVER TANK

This application requires 1,500 cubic feet per minute (cfm) of free air at 90 pounds per square inch-gauge (psig) for 10 minutes each hour. This cycle of 10 minutes at 1,500 cfm and 50 minutes with zero demand repeats hourly. Alternatives obviously are possible: (A) install a 100-psig compressor and standard accessories large enough for the maximum demand requirements; or (B) install a smaller compressor, but for a higher pressure, and store the air in receivers during the off or no-demand period. At least two storage pressures should be considered. In all cases commercial compressor sizes are to be used. For (B1) assume 350 psig and for (B2), 500 psig. (See Table 16-1.)

The 100 psig machine for at least 1,500 cfm output, unit (A) has 1,660 cfm actual capacity and requires 309 brake horsepower (bhp) at full load. The minimum capacity for units B1 and B2 must be calculated, since these can be compressing the full 60 minutes each hour.

$$\text{Total cubic feet per hour} = 1,500 \times 10 = 15,000$$

$$\text{Minimum compressor capacity} = \frac{15,000}{60} = 250 \text{ CFM}$$

Compressor (B1), for 350-psig discharge, is found to have a capacity of 271 cfm and a required brake horsepower of 85. Compressor (B2), for 500 psig, is found to be a standard size with a capacity of 310 cfm and requires 112 bhp. All selections provide some extra capacity for emergency and losses are economical two-stage designs.

Air compressed to 100 psig, to be used at 90 psig, provides no possibility of storage. Storage is practical at the other pressures selected. Compressor units B and C will operate at full load at their rated or lower pressures. Since the receiver pressure will fall during the 10 minutes when air is used faster than the compressor can replenish it, the unit will be operating at full capacity and will supply some of the demand. The full demand need not be stored.

The cubic feet of air to be stored represents the free air, at 14.7 psia, that must be packaged into the receiver above the minimum pressure required by the demand. In this case, the demand pressure is 90 psig, but an allowance for line losses and the necessary reducing valve pressure drop would prevent the use of any air stored below 110 psig. The receiver has a volume of (V) expressed in cubic feet.

$$\text{Useful free air stored} = \frac{V \times \text{Pressure drop}}{14.7}$$

$$V = \frac{\text{Useful free air stored} \times 14.7}{\text{Pressure drop}}$$

This receiver volume may be in one or several tanks, the most economical number being chosen (see Table 16-2).

The final selection can be made only after consideration of the first cost of the compressor, motor, starter, aftercooler, and receiver. The cost of installation, including foundations, piping, and wiring, as well as the operating power cost, must also be considered. In the example, Table 16-3 provides a basic comparison of the three options.

Table 16–1 Comparison of Receiver Options

Unit	B1	B2
Capacity (CFM)	271	310
Demand period (minutes)	10	10
Total cubic feet required	15,000	15,000
Delivered during demand period	2,710	3,100
Cubic feet to be stored	12,290	11,900

Table 16–2 Receiver Volume Required

Unit	B1	B2
Storage pressure (psig)	350	500
Minimum pressure (psig)	110	110
Pressure drop (PSI)	240	390
Free air to be stored (Cu. Ft.)	12,290	11,900
Receiver volume (V)	752	448

Not all problems of this nature will result in the selection of the intermediate storage pressure. Although few will favor the 100-psig level, many will be more economical at the 500-psig level. Experience indicates that there is seldom any gain in using a higher storage pressure than 500 psig, since larger receivers become very expensive above this level and power cost increases.

Air reservoirs are classified as pressure vessels and have to conform to the ASME Pressure Vessel Codes. As such, the following attachments must be fitted:

- Safety valves
- Pressure gauges
- Isolation valves
- Manhole or inspection ports
- Fusible plug

Air reservoirs are designed to receive and store pressurized air. Pressure regulating devices are installed to maintain the pressure within operational limits. When the air reservoir is pressurized to the maximum pressure set point, the pressure regulator causes the air compressor to offload compression by initiating an electrical solenoid valve to use lubricating oil to hydraulically hold open the low-pressure suction valve on the compressor.

As the compressed air is used, the pressure drops in the reservoir until the low-pressure set point is reached. At this point, the pressure-regulated solenoid valve is de-energized. This causes the hydraulic force to drop off on the low-pressure suction valve, restoring it to the full compression cycle.

This cycling process causes drastic variations in noise levels. These noises should not be regarded as problems, unless accompanied by severe knocking or squealing noises. Figure 16-1 shows a typical hydraulic unloader and its location on the compressor.

Table 16–3 Economic Comparison

Unit	A	B1	B2
Pressure (psig)	100	350	500
Installed cost	100%	52%	63%
Fixed charges	100%	52%	63%
Power cost	100%	64%	93%
Oil, water, attendance	100%	100%	100%

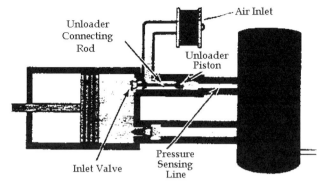

Figure 16–1 Typical hydraulic unloader and its location on the compressor.

17

SAFETY VALVES

All compressed air systems that use a positive displacement compressor must be fitted with a pressure relief or safety valve that will limit the discharge or interstage pressures to a safe maximum limit. Most dynamic compressors must have similar protection because of restrictions placed on casing pressure, power input, and/or keeping out of surge range.

Two types of pressure relief devices are available, *safety valves* and *relief valves*. Although these terms are often used interchangeably, there is a difference between the two. Safety valves are used with gases. The disc overhangs the seat to offer additional thrust area after the initial opening. This fully opens the valve immediately, giving maximum relief capacity. These are often called *pop-off safety valves*.

With relief valves, the disc area exposed to overpressure is the same whether the valve is open or closed. There is a gradual opening, the amount depending upon the degree of overpressure. Relief valves are used with liquids where a relatively small opening will provide pressure relief.

Positive-displacement machines use *safety valves*. There are ASME standards of materials and sizing, and only ASME-stamped valves should be used. The relieving capacity of a given size of safety valve varies materially with the make and design. Care must be taken to ensure proper selection.

An approved safety valve is usually of the "huddling chamber" design. In this valve the static pressure acting on the disc area causes initial opening. As the valve pops, the air space within the huddling chamber between seat and the blowdown ring fills with pressurized air and builds up more pressure on the roof of the disc holder. This temporary pressure increases the upward thrust against the spring, causing the disc and its holder to lift to full pop opening.

After a predetermined pressure drop, which is referred to as blowdown, the valve closes with a positive action by trapping pressurized air on top of the disc holder. The pressure drop is adjusted by raising or lowering the blowdown ring. Raising the ring increases the pressure drop; lowering it decreases the drop.

Most state laws and safe practice require a safety relief valve ahead of the first stop valve in every positive-displacement compressed air system. It is set to release at 1.25 times the normal discharge pressure of the compressor or at the maximum working pressure of the system, whichever is lower. The relief valve piping system sometimes includes a manual vent valve and/or a bypass valve to the suction to facilitate startup and shutdown operations. Quick line sizing equations are (1) line connection, $d/1.75$; (2) bypass, $d/4.5$; (3) vent, $d/6.3$; and (4) relief valve port, $d/9$.

The safety valve is normally situated atop the air reservoir. There must be no restriction on all blow-off points. Compressors can be hazardous to work around because they do have moving parts. Ensure that clothing is kept away from belt drives, couplings, and exposed shafts.

Figure 17-1 illustrates how a safety valve functions.

In addition, high-temperature surfaces around cylinders and discharge piping are exposed. Compressors are notoriously noisy. For this reason, ear protection should be worn. When working around highly pressurized air systems, wear safety glasses and do not search for leaks with bare hands. High-pressure leaks can cause severe friction burns.

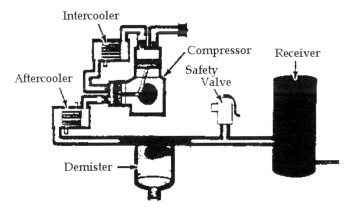

Figure 17–1 Function of a safety valve.

18

COOLERS

The amount of moisture that air can hold is inversely proportional to the pressure of the air. As the pressure of the air increases, the amount of moisture that air can hold decreases. The amount of moisture air can hold is also proportional to the temperature of the air. As the temperature of the air decreases, the amount of moisture it can hold decreases. The pressure change of compressed air is larger than the temperature change of the compressed air. This causes the moisture in the air to condense out of the compressed air. The moisture in compressed air systems can cause serious damage. The condensed moisture could cause corrosion, water hammers, and freeze damage. Therefore, it is important to avoid moisture in compressed-air systems. Coolers are used to address this problem.

Coolers are frequently used on the discharge of a compressor. These are called aftercoolers, and their purpose is to remove the heat generated during the compression of the air. The decrease in temperature promotes the condensation of any moisture present in the compressed air. This moisture is collected in condensate traps that are either automatically or manually drained.

If the compressor is multistage, there may be an intercooler, which is located after the first-stage discharge and second-stage suction. The principle of the intercooler is the same as the principle of the aftercoolers, and the result is drier, cooler compressed air. The structure of the individual cooler depends on the pressure and volume of the air it cools. Figure 18-1 illustrates a typical compressor air cooler.

The combination of drier compressed air (which helps prevent corrosion) and cooler compressed air (which allows more air to be compressed for a set volume) is the reason air coolers are worth the investment.

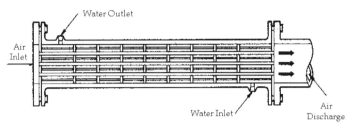

Figure 18–1 Compressor air cooler.

19

VALVES

Pressure Regulators

Pressure regulators are used in the line supplying air-operated devices. These components supply less than full system pressure at a constant outlet pressure unaffected by fluctuations in inlet pressure. Spray guns, air cylinders, hoists, pneumatic screwdrivers, and most other devices having a narrow operating air pressure range benefit from the use of a pressure regulator.

Figure 19-1 shows the essential parts common to most pressure regulators.

The body contains essentially a two-way poppet valve. Two separate forces, acting in opposition, operate this valve. One force is the spring, tending to open the valve; the other force is the diaphragm acting against the spring to permit the valve to close. These two forces working against each other determine the position of the valve and the amount of airflow.

During operation, the spring is set to provide a certain force against the valve. The valve is opened to supply pressure and allows air to flow. Flow through the regulator and downstream resistance cause the outlet pressure to increase. This increase acts against the underside of the diaphragm. The pressure will rise until the diaphragm exerts sufficient force to just balance the spring force. At this point the inlet pressure will close the valve.

As the air is used, the outlet pressure falls and the spring can again overpower the diaphragm until the valve opens enough to restore the desired outlet pressure setting. Thus, a nearly constant pressure is maintained, regardless of airflow, up to the capacity of the regulator.

In all pressure regulators, as the flow is increased the outlet pressure drops below the original pressure setting. This deviation between the actual and set pressure is called

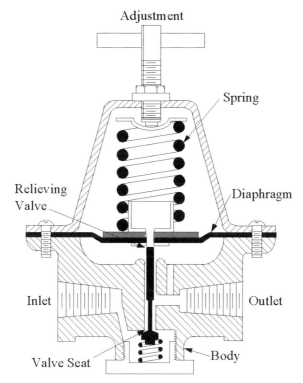

Figure 19–1 Pressure regulator.

the regulator "droop." There are ways of compensating the regulator to minimize this "droop," but it will always be present to some extent.

The illustrated regulator is of the self-relieving type. It consists of a port through the diaphragm and is normally closed by the valve stem. The spring holds the diaphragm against the stem and closes the small valve. If the outlet pressure rises over the setting, it lifts the diaphragm off the stem. This causes the port to open and allows excess air to escape, reducing the pressure. Relieving regulators have the advantage of venting excess air when the regulator setting is decreased. If pressure should rise in the outlet pressure circuit, in a nonrelieving-type regulator, the regulator will not control it.

FLOWMETERS

Airflow must be monitored regularly for a variety of reasons. Flow measurement is required to determine the air consumption of pneumatic tools and devices, actual compressor output, and the pressure drop characteristics of valves and other components.

The measurement of flow in pneumatic systems is much more difficult than it is in hydraulic systems. Hydraulic flow is easily determined by catching and measuring the volume of liquid that flows in a known period of time. For small airflows this can also be done with a water-displacement chamber, but it is not usually convenient. So, indirect methods are used. The airflow measuring devices include rotometers, orifice meters, pitot tubes, and a variety of other special devices. For industrial purposes, airflow is commonly stated in terms of cubic feet per minute (cfm) for large flows, and cubic feet per hour (cfh) for small flows. Flow may be measured at any desired pressure, but is usually stated in terms of flow at standard conditions (14.7 psia and 70°F).

The most common measuring device for airflow is the *rotometer*. This device consists of a tapered vertical tube with a ball or other type of "float" in it. The airflow lifts the float upward until it reaches a point where clearance is sufficient to allow the flow to pass. A scale is provided on the tube so that the flow is indicated by the height of the ball against the scale markers.

The flowmeter is calibrated for atmospheric-pressure operation. At higher pressures the air is denser. Because the air is denser, the flow readings indicate less than the actual flow. To compensate for this, corrections must be made by using chart in Figure 19-2.

For highest accuracy, a correction should be made for air temperature also, but this can be neglected at normal temperatures.

To use the correction chart, do the following:

1. Draw a line horizontally from the observed reading to where it intersects the diagonal line for the actual flowmeter pressure.

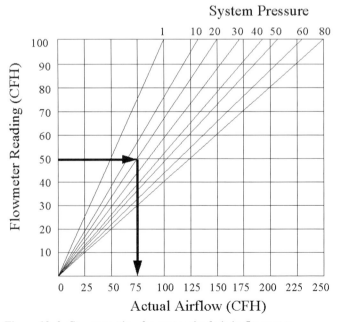

Figure 19–2 Compensation for pressurized air in flowmeter.

2. From that intersection, draw a vertical line down to read actual flow at the bottom of the chart.
3. The example shown in the graph is a flowmeter reading of 50 cfh, at 20 psig. The actual airflow is 75 cubic feet per hour (cfh).

FLOW CONTROL VALVES

Flow control valves come in all shapes, sizes, and designs. Their basic function, however, is the same—to control flow of air. Flow control valves for hydraulic systems (liquids under pressure) are of the same basic design. A typical example of a flow control valve is the simple water faucet installed in homes.

Globe valves and needle valves are standard designs used for flow control. Unidirectional flow control valves control the flow in one direction but permit free flow in the other direction. Pressure-compensated flow control valves are also manufactured. These valves control the amount of flow and will maintain a constant flow at different pressures. These valves are ideal for some applications but should be used only when required because of their higher cost.

The check valve is another type of flow control valve. The function of a check valve is to permit flow in only one direction. A very common function of flow control valves is to control the speed of cylinders and air motors. The speed of cylinders or air motors depends on the amount of air, which can be controlled by flow control valves.

DIRECTIONAL CONTROL VALVES

Directional control valves are used to control the operation of pneumatic cylinders and air motors. The most commonly used have moving inner spools or piston slide valves. Moving the spool changes the airflow patterns by redirecting the output to perform different functions. Figure 19-3 depicts the simplified flow pattern through the valve.

The simplest type of valve is a two-way valve. With this type of valve there are two ports (ports are simply openings for air to pass through). In one position this valve

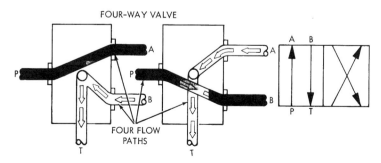

Figure 19–3 Directional control valve: four-way, two-position.

will allow air passage. In the other position it allows no flow. Figure 19-4 shows the function of a two-way valve.

Three-way valves usually have three ports. Pressure goes in one port and is shifted between the other two ports by the spool or piston slide valve. This type of valve is typically used when high cylinder speeds are needed or spring return cylinders are used. Figure 19-5 shows the function and flow through three-way valves.

Four-way valves are probably the most common type of directional control valve. Their porting allows cylinders and air motors to be reversed. Some four-way valves have four ports (only one exhaust) and some have five ports (two exhausts). An advantage of the five-port valve is its capability to control exhaust flow from two exhaust ports. This allows the extension and retraction speed of a cylinder to be controlled independently. Figure 19-6 shows the functions of the five different ports.

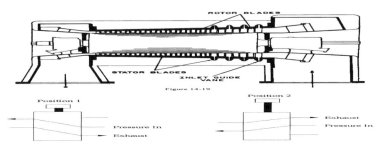

Figure 19–4 Directional valve: two-way, normally open.

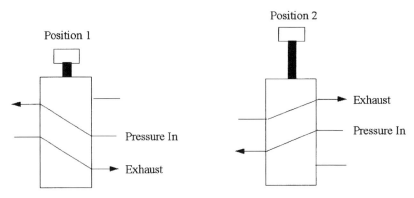

Figure 19–5 Directional valve: three-way, normally closed.

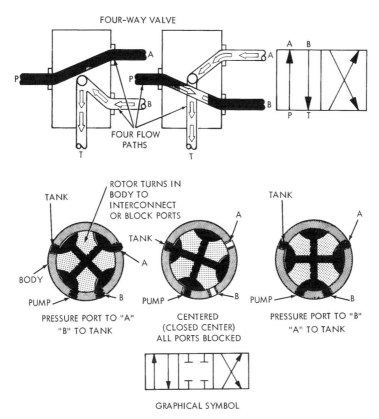

Figure 19–6 Directional valve: four-way, two-position.

20

ACTUATORS

PNEUMATIC CYLINDERS

Cylinders are the muscles of pneumatic systems. The valves and other control devices simply provide the correct amount and pressure of air at the required time to the cylinders. The cylinders then produce linear motion. This linear motion is ideally suited for opening and shutting large valves and moving the large loads that are commonly found in power stations and other industrial complexes.

A pneumatic cylinder basically consists of a cylindrical body, a piston rod, a closed end with an outlet port, and the other end with a cylinder head through which the piston rod slides in and out. A rod seal in the cylinder head prevents air blow-through when the cylinder is pressurized. Figure 20-1 shows a typical pneumatic cylinder and its components.

Cylinders can be found in many different sizes and can vary in stroke output depending upon the designed application. Some are constructed for use in very low-pressure systems and work capacities; others perform incredible load-moving feats utilizing

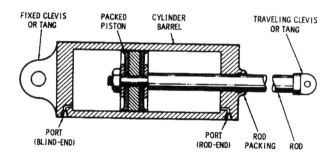

Figure 20–1 Pneumatic cylinder and its components.

extremely high pressures. Cylinder designs come in many sizes, and are usually described in the following manner:

2 inches by 10 inches

The "2 inches" describes the *diameter* of the cylinder and the "10 inches" describes the *stroke* of the cylinder (how far the piston rod moves). Cylinders vary in design, but the following designs are standard

Single-Acting Cylinders

In this cylinder design (Figure 20-2), there is only one pressure port. Instead of switching pressure to the opposite side of the piston head to reverse the cylinder, a spring inside the cylinder (fitted between the backside of the piston head and the cylinder head enclosure), or the load itself, reverses the cylinder's movement.

Double-Acting Cylinder

This type of cylinder has two pressure ports (Figure 20-3). Either of these ports can be pressurized to cause the piston rod to move in either direction. This design is the most common one in use throughout industry.

Double-End Rod Cylinder

This type of cylinder utilizes two piston rods and two ports. A piston rod extends from both ends of the cylinder and produces a force in both directions.

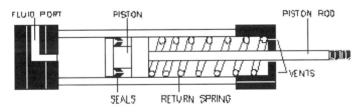

Figure 20–2 Single-acting cylinder.

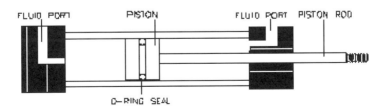

Figure 20–3 Double-acting cylinder.

PNEUMATIC–MECHANICAL SYSTEMS

Almost all pneumatic cylinders operate some kind of mechanical device. The most common type of device is the lever or lever system.

The change in *force* resulting from a mechanical system is called *mechanical advantage*. The use of a lever system typically provides an increase or decrease in mechanical advantage. It is very important to realize that the same amount of work is being done regardless of mechanical advantage.

Figure 20-4 shows how a lever can be made to work to our advantage.

The components of a lever system are a lever, lever arm, force, and load. Another important part of the system is the fulcrum. The fulcrum is the point about which the lever rotates. The distance from the fulcrum to the load multiplied by the amount of load will be equal to the distance from the fulcrum to the force multiplied by the amount of force required to move the load. Attaching a cylinder to supply the force and utilizing the proper distances from the fulcrum allows cylinders to move very large loads.

PRESSURE, FORCE, AND AREA

The relationship between pressure, force, and area is one of the most important concepts in pneumatics. It can be used to calculate the forces in the cylinders, tubing, and other pneumatic equipment. The relationship is as follows:

$$\text{Pressure} = \frac{\text{Force}}{\text{Area}}$$

This can be arranged as:

$$\text{Force} = \text{Area} \times \text{Pressure}$$

or

$$\text{Area} = \frac{\text{Force}}{\text{Pressure}}$$

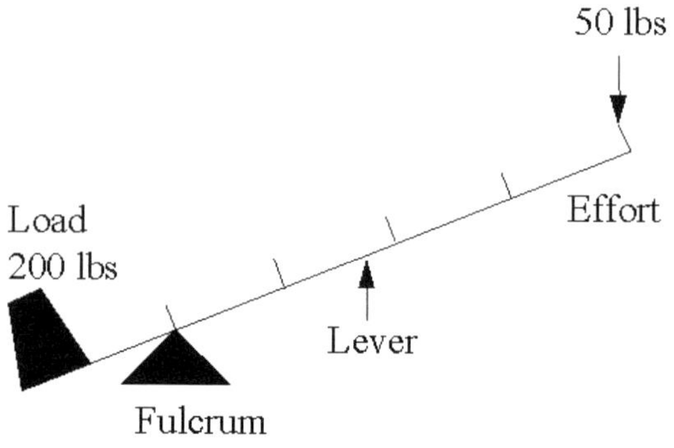

Figure 20–4 Fulcrum lever.

The formula for finding the cross-sectional area of a cylinder is:

$$\text{Area} = \pi \times \text{Radius}^2$$

or

$$\text{Area} = \frac{\pi \, \text{Diameter}^2}{4}$$

Example: To find the cross-sectional area of a 5 × 10 inch cylinder:

$$\text{Area} = \pi \, R^2$$

$$\text{Area} = 3.1416 \times 2.5^2$$

$$\text{Area} = 3.1416 \times 6.25$$

$$\text{Area} = 19.635 \text{ inches}^2$$

The pressure refers to the actual system pressure in pounds per square inch. Typically this will be the pressure indicated on the manifold gauge and is the gauge pressure (psig). The force developed by the cylinder is a function of the pressure and the area of the cylinder. The force in most cases is the product of the gauge pressure (psig) and the area of the cylinder (inches^2).

MULTIPLE CYLINDER CIRCUITS

In the majority of pneumatic applications, more than one cylinder is used in a circuit. Depending upon the application, the cylinders may be required to operate at different time intervals and be expected to exert different forces. There are a wide variety of pneumatic circuit designs that enable this type of operation to meet the requirements of an application.

A typical application using more than one cylinder is a pneumatic-powered robotic arm. A robotic arm uses two or more cylinders, which must be synchronized with each other to perform precise movements for exact positioning (Figure 20-5).

CYLINDER AIR CONSUMPTION

The purpose of estimating the air consumption of a cylinder is usually to find the horsepower capacity that must be available from the air compressor to operate the cylinder on a continuous cycling application. Air consumption can be estimated from Table 20-1. The consumption can then be converted into compressor horsepower.

Figures in the body of the table are air consumption for cylinders with standard diameter piston rods. The saving of air for cylinders with larger diameter rods is negligible

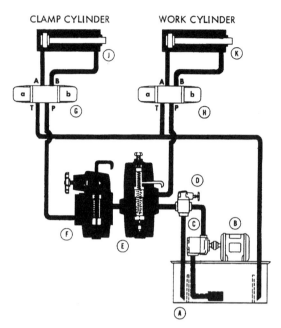

Figure 20–5 Layout of multiple cylinders.

for most calculations. Air consumption was calculated assuming the cylinder piston will be allowed to stall, at least momentarily, at each end of its stroke, giving it time to fill up with air to the pressure regulator setting. If reversed at either end of its stroke before full stall occurs, air consumption will be less than that shown in the table.

The first step in the calculation is to be sure that the bore size of the selected cylinder will just balance the load at a pressure of 75 percent or less of the maximum pressure available to the system. This leaves about 25 percent of available pressure, which can be used to overcome flow losses through piping and valving. This surplus pressure must be available or the cylinder cannot travel at normal speed.

Determine the exact air pressure needed to just balance the load resistance. Add about 25 percent for flow losses and set the system regulator to this pressure. This is also the pressure figure, which should be used when going into the table. After determining the regulator pressure, go into the proper column of the table. The figures shown in the table are the air consumption for a 1-inch stroke, forward and return. Take this figure and multiply by the number of inches of actual stroke and by the number of complete cycles forward and back that the cylinder is expected to make in 1 minute. This gives the *standard cubic feet per minute* (scfm) for the application.

CONVERTING SCFM INTO COMPRESSOR HORSEPOWER

Compression of air is an inefficient process because part of the energy is lost as heat of compression and can never be recovered. Overcompressing the air and then reduc-

Table 20–1 Cylinder Air Consumption per 1-inch Stroke, Forward and Return

Cylinder bore	60 PSI	70 PSI	80 PSI	90 PSI	100 PSI	110 PSI	120 PSI	130 PSI	140 PSI	150 PSI
1.50	.009	.010	.012	.013	.015	.016	.017	.018	.020	.021
2.00	.018	.020	.022	.025	.027	.029	.032	.034	.035	.039
2.50	.028	.032	.035	.039	.043	.047	.050	.054	.058	.062
3.00	.039	.044	.050	.055	.060	.066	.070	.076	.081	.087
3.25	.046	.053	.059	.065	.071	.078	.084	.090	.096	.102
4.00	.072	.081	.091	.100	.110	.119	.129	.139	.148	.158
5.00	.113	.128	.143	.159	.174	.189	.204	.219	.234	.249
6.00	.162	.184	.205	.227	.249	.270	.292	.314	.335	.357
8.00	.291	.330	.369	.408	.447	.486	.525	.564	.602	.642
10.0	.455	.516	.576	.637	.698	.759	.820	.881	.940	1.00
12.0	.656	.744	.831	.919	1.01	1.09	1.18	1.27	1.36	1.45
14.0	.890	1.01	1.13	1.25	1.37	1.49	1.61	1.72	1.84	1.96

ing it to a lower pressure through a regulator increases the system losses. The amount of this loss is nearly impossible to calculate, but on the average system may amount to 5 or 10 percent. Also, there is a small loss due to flow resistance through the regulator. After finding the scfm to operate the cylinder, refer to Table 20-2. Convert into horsepower according to the kind of compressor used. Add 5 to 10 percent for the miscellaneous losses described above. This should be very close to the actual horsepower capacity needed.

One important use for this table is to estimate the compressor horsepower capacity needed to operate an air cylinder. The scfm required by the cylinder under stated operating conditions must first be calculated by the method discussed earlier. Then the appropriate column in the table can be used to convert scfm into horsepower. For example, if cylinder consumption has been calculated to be 24 scfm, and if the compressor is a two-stage model, the horsepower needed at 90 psig will be:

$$HP = 24 \times .156 = 3.74$$

AIR CYLINDER FORCE

In Table 20-3, cylinder forces are shown in pounds for both extension and retraction. Lines in bold type show extension forces, using the full piston area. Lines in italic type show retraction forces with various size piston rods. Remember that force values are theoretical, derived by calculation.

Pressures along the top of the chart do not represent air supply pressure; they are differential pressures across the two cylinder ports. In practice, the air supply line must supply another 5 percent of pressure to make up for cylinder loss, and must supply an estimated 25 to 50 percent additional pressure to make up for flow losses in lines and valving so the cylinder will have sufficient travel speed.

Table 20–2 Horsepower for Compressing Air (Efficiency of All Compressors Is Assumed to Be 85 Percent)

1-Stage compressor		2-Stage compressor		3-Stage compressor	
psig	HP0	psig	HP0	psig	HP0
5	.021	50	.116	100	.159
10	.040	60	.128	150	.190
15	.056	70	.138	200	.212
20	.067	80	.148	250	.230
25	.079	90	.156	300	.245
30	.095	100	.164	350	.258
35	.099	110	.171	400	.269
40	.107	120	.178	450	.279
45	.116	130	.185	500	.289
50	.123	140	.190	550	.297
55	.130	150	.198	600	.305
60	.136	160	.201	650	.311
65	.143	170	.206	700	.317
70	.148	180	.211	750	.323
75	.155	190	.216	800	.329
80	.160	200	.220	850	.335
85	.166	210	.224	900	.340
90	.170	220	.228	950	.345
95	.175	230	.232	1,000	.350
100	.179	240	.236	1,050	.354
110	.188	250	.239	1,100	.358
120	.196	260	.243	1,150	.362
130	.204	270	.246	1,200	.366
140	.211	280	.250	1,250	.370
150	.218	290	.253	1,300	.374
160	.225	300	.255	1,350	.378
170	.232	350	.269	1,400	.380
180	.239	400	.282	1,450	.383
190	.244	450	.293	1,500	.386
200	.250	500	.303	1,550	.390

For good design and highest circuit efficiency, open the cylinder speed control valves as wide as practical and reduce the pressure regulator setting to as low a pressure as will give satisfactory cylinder force and speed. For pressures not shown, use the effective areas in the third column as power factors. Multiply effective area times differential pressure to obtain theoretical cylinder force.

Table 20–3 Air Cylinder Extension and Retraction Forces

Piston diameter (inches)	Rod diameter (inches)	Effective area (inches²)	60 PSI	70 PSI	80 PSI	90 PSI	100 PSI	110 PSI	120 PSI	130 PSI
1-1/2	None	1.77	106	124	142	159	177	195	212	230
	5/8	1.46	88	102	117	132	146	161	176	190
	1	.985	59	69	79	89	98	108	118	128
1-3/4	None	2.41	144	168	192	216	241	265	289	313
	5/8	2.10	126	147	168	189	210	231	252	273
	1-1/4	1.18	71	83	95	106	118	130	142	154
2	None	3.14	188	220	251	283	314	345	377	408
	5/8	2.83	170	198	227	255	283	312	340	368
	1	2.35	141	165	188	212	235	259	283	306
2-1/2	None	4.91	295	344	393	442	491	540	589	638
	5/8	4.60	276	322	368	414	460	506	552	598
	1	4.12	247	289	330	371	412	454	495	536
	1-3/8	3.43	206	240	274	308	343	377	411	445
3	None	7.07	424	495	565	636	707	778	848	919
	1	6.28	3.77	440	503	565	628	691	754	817
	1-3/4	4.66	2.80	326	373	420	466	513	560	606
3-1/4	None	8.30	498	581	664	747	830	913	996	1,079
	1	7.51	451	526	601	676	751	827	902	977
	1-3/8	6.82	409	477	545	613	681	750	818	886
	1-3/4	5.89	354	413	472	531	589	648	707	766
3-1/2	None	9.62	577	674	770	866	962	1,058	1,155	1,251
	1	8.84	530	618	707	795	884	972	1,060	1,149
4	None	12.57	754	880	1,006	1,131	1,257	1,283	1,508	1,634
	1	11.78	707	825	943	1,061	1,178	1,296	1,415	1,532
	1-3/8	11.09	665	776	887	998	1,109	1,219	1,330	1,441
	1-3/4	10.16	610	712	813	915	1,016	1,118	1,220	1,321
5	None	19.64	1,178	1,375	1,571	1,768	1,964	2,160	2,357	2,553
	1	18.85	1,131	1,320	1,508	1,697	1,885	2,074	2,263	2,451
	1-3/8	18.16	1,089	1,271	1,452	1,634	1,816	1,997	2,179	2,360
6	None	28.27	1,696	1,979	2,262	2,544	2,827	3,110	3,392	3,675
	1-3/8	26.79	1,607	1,875	2,143	2,411	2,679	2,946	3,214	3,482
	1-3/4	25.90	1,552	1,811	2,069	2,328	2,586	2,845	3,104	3,362
7	None	38.49	2,309	2,694	3,079	3,464	3,849	4,234	4,619	5,004
	1-3/8	37.01	2,220	2,590	2,960	3,331	3,701	4,071	4,441	4,811
8	None	50.27	3,016	3,519	4,022	4,524	5,027	5,530	6,032	6,535
	1-3/8	48.79	2,927	3,415	3,903	4,391	4,879	5,366	5,854	6,342
	1-3/4	47.90	2,872	3,351	3,829	4,308	4,786	5,265	5,744	6,222
10	None	78.54	4,712	5,498	6,283	7,069	7,854	8,639	9,425	10,210
	1-3/4	76.14	4,568	5,329	6,091	6,852	7,614	8,375	9,136	9,898
	2	75.40	4,524	5,278	6,032	6,786	7,540	8,294	9,048	9,802
12	None	113.1	6,786	7,917	9,048	10,179	11,310	12,441	13,572	14,703
	2	110.0	6,598	7,697	8,797	9,896	10,996	12,095	13,195	14,295
	2-1/2	108.2	6,491	7,573	8,655	9,737	10,819	11,901	12,983	14,075
14	None	153.9	9,234	10,773	12,312	13,851	15,390	16,929	18,468	20,007
	2-1/2	149.0	8,939	10,429	11,919	13,409	14,899	16,389	17,879	19,369
	3	146.8	8,810	10,278	11,747	13,215	14,683	16,151	17,620	19,088

21

TROUBLESHOOTING PNEUMATIC CIRCUITS

Troubleshooting is the process of solving problems in systems. Effective troubleshooting requires a good understanding of pneumatics and the flow circuit in trouble. Safety is an important consideration when troubleshooting circuits. Very often high pressures and large loads are involved. Always ensure that you, and the machine, are protected from injury that could occur during analysis, testing, or repair. The following is a general troubleshooting procedure.

TROUBLESHOOTING PNEUMATIC CIRCUITS

The first step is to gain a practical understanding of the pneumatic system. This understanding must begin with a design review. Study the circuit and its diagram to be sure you understand the function of the circuit and the air flow paths under all operating conditions.

After studying the circuit and observing the problem, you should have an indication of where the problem is occurring. If the problem cannot be pinpointed, the following steps outline an overall troubleshooting program.

First, check for obvious problems, such as inadequate air pressure, large leaks, empty lubricators, clogged filters, or plugged or chinked air lines, that would result in system failure or reduced efficiency. Then, check the control valves to ensure that they are shifting properly. Valves are usually very durable, but they can cause problems in the long run or if they are improperly maintained. If electrical or mechanical actuators are used, these should also be inspected to ensure proper operation.

Actuators, such as cylinders and air motors, should then be tested for proper function. Again, it should be stressed that an understanding of the concepts of pneumatics and a thorough knowledge of the circuit are the most important aspects of successfully troubleshooting a circuit.

In many cases, these simple steps can be used to isolate problems that result in partial or complete failure of pneumatic circuits. In those cases where these steps fail to find the problem, each of the system's components must be tested for proper operation.

PNEUMATIC SYSTEM COMPONENT FAILURE MODES

Each of the components that make up a pneumatic system has inherent design strengths and weaknesses. A thorough understanding of these inherent characteristics will be helpful in your troubleshooting efforts.

Compressors

Compressors can be divided into three classifications: centrifugal, rotary, and reciprocating. This section identifies the common failure modes for each.

Centrifugal Compressors

The operating dynamics of centrifugal compressors are the same as for other centrifugal machine trains. The dominant forces and vibration profiles are typically identical to pumps or fans. However, the effects of variable load and other process variables (e.g., temperatures, inlet/discharge pressure) are more pronounced than in other rotating machines. Table 21-1 identifies the common failure modes for centrifugal compressors.

Aerodynamic instability is the most common failure mode for centrifugal compressors. Variable demand and restrictions of the inlet air flow are common sources of this instability. Even slight variations can cause dramatic changes in the operating stability of the compressor.

Entrained liquids and solids can also affect operating life. When dirty air must be handled, open-type impellers should be used. An open design provides the ability to handle a moderate amount of dirt or other solids in the inlet air supply. However, inlet filters are recommended for all applications, and controlled liquid injection for cleaning and cooling should be considered during the design process.

Rotary-Type Positive-Displacement Compressors

Table 21-2 defines the common failure modes of rotary-type positive-displacement compressors. This type of compressor can be grouped into two types: sliding vane and rotary screw.

Table 21–1 Common Failure Modes of Centrifugal Compressors

THE CAUSES	Excessive Vibration	Compressor Surges	Loss of Discharge Pressure	Low Lube Oil Pressure	Excessive Bearing Oil Drain Temp.	Units Do Not Stay in Alignment	Persistent Unloading	Water in Lube Oil	Motor Trips
THE PROBLEM									
Bearing Lube Oil Orifice Missing or Plugged				●					
Bent Rotor (Caused by Uneven Heating and Cooling)	●						●		
Build-up of Deposits on Diffuser		●							
Build-up of Deposits on Rotor	●	●							
Change in System Resistance		●							●
Clogged Oil Strainer/Filter				●					
Compressor Not Up To Speed			●						
Condensate in Oil Reservoir								●	
Damaged Rotor	●								
Dry Gear Coupling	●								
Excessive Bearing Clearance	●								
Excessive Inlet Temperature				●					
Failure of Both Main and Auxiliary Oil Pumps				●					
Faulty Temperature Gauge or Switch				●	●				●
Improperly Assembled Parts	●						●		●
Incorrect Pressure Control Valve Setting				●					
Insufficient Flow		●							
Leak In Discharge Piping			●						
Leak In Lube Oil Cooler Tubes or Tube Sheet								●	
Leak in Oil Pump Suction Piping				●					
Liquid "Slugging"	●						●		
Loose or Broken Bolting	●								
Loose Rotor Parts	●								
Oil Leakage				●					
Oil Pump Suction Plugged				●					
Oil Reservoir Low Level				●					
Operating at Low Speed w/o Auxiliary Oil Pump				●					
Operating in Critical Speed Range	●								
Operating in Surge Region	●								
Piping Strain	●					●	●	●	●
Poor Oil Condition					●				
Relief Valve Improperly Set or Stuck Open				●					
Rotor Imbalance	●						●		
Rough Rotor Shaft Journal Surface					●		●		●
Shaft Misalignment	●					●			
Sympathetic Vibration	●						●	●	
Vibration					●				
Warped Foundation or Baseplate							●		●
Wiped or Damaged Bearings					●				●
Worn or Damaged Coupling	●								

Sliding-Vane Compressors

Sliding-vane compressors have the same failure modes as vane-type pumps. The dominant components in their vibration profile are running speed, vane-pass frequency, and bearing-rotation frequencies. In normal operation, the dominant energy is at the shaft's running speed. The other frequency components are at much lower energy levels. Common failures of this type of compressor occur with shaft seals, vanes, and bearings.

Table 21–2 Common Failure Modes of Rotary-Type Positive-Displacement Compressors

THE CAUSES	No Air/Gas Delivery	Insufficient Discharge Pressure	Insufficient Capacity	Excessive Wear	Excessive Heat	Excessive Vibration and Noise	Excessive Power Demand	Motor Trips	Elevated Motor Temperature	Elevated Air/Gas Temperature
Air Leakage Into Suction Piping or Shaft Seal		●	●			●				
Coupling Misaligned				●	●	●	●		●	
Excessive Discharge Pressure				●	●		●	●	●	●
Excessive Inlet Temperature/Moisture				●						
Insufficient Suction Air/Gas Supply		●	●	●		●			●	
Internal Component Wear	●	●	●							
Motor or Driver Failure	●									
Pipe Strain on Compressor Casing				●	●	●	●		●	
Relief Valve Stuck Open or Set Wrong		●	●							
Rotating Element Binding				●	●	●	●	●	●	
Solids or Dirt in Inlet Air/Gas Supply				●						
Speed Too Low		●	●						●	
Suction Filter or Strainer Clogged	●	●	●			●			●	
Wrong Direction of Rotation	●	●							●	

Shaft seals. Leakage through the shaft's seals should be checked visually once a week or as part of every data-acquisition route. Leakage may not be apparent from the outside of the gland. If the fluid is removed through a vent, the discharge should be configured for easy inspection. Generally, more leakage than normal is the signal to replace a seal. Under good conditions, they have a normal life of 10,000 to 15,000 hours and should routinely be replaced when this service life has been reached.

Vanes. Vanes wear continuously on their outer edges and, to some degree, on the faces that slide in and out of the slots. The vane material is affected somewhat by prolonged heat, which causes gradual deterioration. Typical life expectancy of vanes in 100-psig service is about 16,000 hours of operation. For low-pressure applications, life may reach 32,000 hours.

Replacing vanes before they break is extremely important. Breakage during operation can severely damage the compressor, which requires a complete overhaul and realignment of heads and clearances.

Bearings. In normal service, bearings have a relatively long life. Replacement after about 6 years of operation is generally recommended. Bearing defects are usually displayed in the same manner in a vibration profile as for any rotating machine train. Inner and outer race defects are the dominant failure modes, but roller spin also may contribute to the failure.

Rotary Screw Compressors

The most common reason for compressor failure or component damage is process instability. Rotary screw compressors are designed to deliver a constant volume and pressure of air or gas. These units are extremely susceptible to any change in either inlet or discharge conditions. A slight variation in pressure, temperature, or volume can result in instantaneous failure. The following are used as indices of instability and potential problems: rotor mesh, axial movement, thrust bearings, and gear mesh.

Rotor mesh. In normal operation, the vibration energy generated by male and female rotor meshing is very low. As the process becomes unstable, the energy due to the rotor-meshing frequency increases, with both the amplitude of the meshing frequency and the width of the peak increasing. In addition, the noise floor surrounding the meshing frequency becomes more pronounced. This white noise is similar to that observed in a cavitating pump or unstable fan.

Axial movement. The normal tendency of the rotors and helical timing gears is to generate axial shaft movement, or thrusting. However, the extremely tight clearances between the male and female rotors do not tolerate any excessive axial movement, and therefore, axial movement should be a primary monitoring parameter. Axial measurements are needed from both rotor assemblies. If there is any increase in the vibration amplitude of these measurements, it is highly probable that the compressor will fail.

Thrust bearings. Although process instability can affect both the fixed and float bearings, the thrust bearing is more likely to show early degradation as a result of process instability or abnormal compressor dynamics. Therefore, these bearings should be monitored closely and any degradation or hint of excessive axial clearance should be corrected immediately.

Gear mesh. The gear mesh vibration profile also provides an indication of prolonged compressor instability. Deflection of the rotor shafts changes the wear pattern on the helical gear sets. This change in pattern increases the backlash in the gear mesh, results in higher vibration levels, and increases thrusting.

Reciprocating Positive-Displacement Compressors

Reciprocating compressors have a history of chronic failures that include valves, lubrication system, pulsation, and imbalance. Table 21-3 identifies common failure modes and causes for this type of compressor.

Like all reciprocating machines, reciprocating compressors normally generate higher levels of vibration than centrifugal machines. In part, the increased level of vibration is due to the impact as each piston reaches top dead center and bottom dead center of its stroke. The energy levels also are influenced by the unbalanced forces generated by non-opposed pistons and looseness in the piston rods, wrist pins, and journals of the

Table 21–3 Common Failure Modes of Reciprocating Compressor

THE PROBLEM

THE CAUSES	Air Discharge Temperature Above Normal	Carbonaceous Deposits Abnormal	Compressor Fails to Start	Compressor Fails To Unload	Compressor Noisy or Knocks	Compressor Parts Overheat	Crankcase Oil Pressure Low	Crankcase Water Accumulation	Delivery Less Than Rated Capacity	Discharge Pressure Below Normal	Excessive Compressor Vibration	Intercooler Pressure Above Normal	Intercooler Pressure Below Normal	Intercooler Safety Valve Pops	Motor Over-heating	Oil Pumping Excessive (Single-acting Compressor)	Operating Cycle Abnormally Long	Outlet Water Temperature Above Normal	Piston Ring, Piston, Cylinder Wear Excessive	Piston Rod or Packing Wear Excessive	Receiver Pressure Above Normal	Receiver Safety Valve Pops	Starts Too Often	Valve wear and breakage normal
Air Discharge Temperature Too High		●															●							
Air Filter Defective		●																	●	●				●
Air Flow to Fan Blocked	●	●			●																			
Air Leak into Pump Suction								●																
Ambient Temperature Too High	●	●			●								●											●
Assembly Incorrect																								●
Bearings Need Adjustment or Renewal					●	●	●						●											
Belts Slipping					●				●	●														
Belts Too Tight				●		●							●											
Centrifugal Pilot Valve Leaks					●																			
Check or Discharge Valve Defective					●																			
Control Air Filter, Strainer Clogged				●																				
Control Air Line Clogged																						●	●	
Control Air Pipe Leaks																					●	●	●	
Crankcase Oil Pressure Too High													●											
Crankshaft End Play Too Great					●																			
Cylinder, Head, Cooler Dirty	●	●																						
Cylinder, Head, Intercooler Dirty					●													●						
Cylinder (Piston) Worn or Scored	●	●			●	●			●	●	●	●H	●L	●H	●L	●	●		●H	●H				
Detergent Oil Being Used (3)								●																
Demand Too Steady (2)																							●	
Dirt, Rust Entering Cylinder		●																			●	●		●

Table 21–3 Continued.

THE PROBLEM

THE CAUSES	Air Discharge Temperature Above Normal	Carbonaceous Deposits Abnormal	Compressor Fails to Start	Compressor Fails To Unload	Compressor Noisy or Knocks	Compressor Parts Overheat	Crankcase Oil Pressure Low	Crankcase Water Accumulation	Delivery Less Than Rated Capacity	Discharge Pressure Below Normal	Excessive Compressor Vibration	Intercooler Pressure Above Normal	Intercooler Pressure Below Normal	Intercooler Safety Valve Pops	Motor Over-heating	Oil Pumping Excessive (Single-acting Compressor)	Operating Cycle Abnormally Long	Outlet Water Temperature Above Normal	Piston Ring, Piston, Cylinder Wear Excessive	Piston Rod or Packing Wear Excessive	Receiver Pressure Above Normal	Receiver Safety Valve Pops	Starts Too Often	Valve wear and breakage normal
Discharge Line Restricted	●														●									
Discharge Pressure Above Rating	●	●			●	●			●			●	●		●	●		●	●	●	●	●		
Electrical Conditions Wrong			●												●									
Excessive Number of Starts															●									
Excitation Inadequate			●												●									
Foundation Bolts Loose					●						●													
Foundation Too Small											●													
Foundation Uneven—Unit Rocks					●						●													
Fuses Blown			●																					
Gaskets Leak	●	●			●	●			●	●		●H	●L	●H	●L	●			●H	●H				
Gauge Defective								●		●	●	●	●										●	
Gear Pump Worn/Defective							●																	
Grout, Improperly Placed											●													
Intake Filter Clogged	●				●	●			●	●			●			●	●	●						
Intake Pipe Restricted, Too Small, Too Long	●				●	●			●	●			●			●	●	●						
Intercooler, Drain More Often								●																
Intercooler Leaks												●												
Intercooler Passages Clogged												●	●											
Intercooler Pressure Too High																		●						
Intercooler Vibrating					●																			
Leveling Wedges Left Under Compressor											●													
Liquid carry-over					●			●											●	●	●			●

Table 21–3 Continued.

THE PROBLEM

THE CAUSES	Air Discharge Temperature Above Normal	Carbonaceous Deposits Abnormal	Compressor Fails to Start	Compressor Fails To Unload	Compressor Noisy or Knocks	Compressor Parts Overheat	Crankcase Oil Pressure Low	Crankcase Water Accumulation	Delivery Less Than Rated Capacity	Discharge Pressure Below Normal	Excessive Compressor Vibration	Intercooler Pressure Above Normal	Intercooler Pressure Below Normal	Intercooler Safety Valve Pops	Motor Over-heating	Oil Pumping Excessive (Single-acting Compressor)	Operating Cycle Abnormally Long	Outlet Water Temperature Above Normal	Piston Ring, Piston, Cylinder Wear Excessive	Piston Rod or Packing Wear Excessive	Receiver Pressure Above Normal	Receiver Safety Valve Pops	Starts Too Often	Valve wear and breakage normal
Location Too Humid and Damp								●																
Low Oil Pressure Relay Open			●																					
Lubrication Inadequate	●				●	●									●			●	●	●				●
Motor Overload Relay Tripped			●																					
Motor Rotor Loose on Shaft					●						●													
Motor Too Small			●												●									
New Valve on Worn Seat																								●
"Off" Time Insufficient	●	●			●																			
Oil Feed Excessive		●			●															●				●
Oil Filter or Strainer Clogged							●																	
Oil Level Too High	●	●			●	●											●							
Oil Level Too Low						●	●																	
Oil Relief Valve Defective							●																	
Oil Viscosity Incorrect		●			●	●	●								●	●		●	●					●
Oil Wrong Type																●								
Packing Rings Worn, Stuck, Broken																				●				
Piping Improperly Supported											●													
Piston or Piston Nut Loose					●																			
Piston or Ring Drain Hole Clogged																●								
Piston Ring Gaps Not Staggered																●								
Piston Rings Worn, Broken, or Stuck	●	●			●	●		●	●	●	●	●H	●L	●H	●L	●	●		●H	●H				
Piston-to-head Clearance Too Small					●																			

Table 21–3 Continued.

THE PROBLEM

THE CAUSES	Air Discharge Temperature Above Normal	Carbonaceous Deposits Abnormal	Compressor Fails to Start	Compressor Fails To Unload	Compressor Noisy or Knocks	Compressor Parts Overheat	Crankcase Oil Pressure Low	Crankcase Water Accumulation	Delivery Less Than Rated Capacity	Discharge Pressure Below Normal	Excessive Compressor Vibration	Intercooler Pressure Above Normal	Intercooler Pressure Below Normal	Intercooler Safety Valve Pops	Motor Over-heating	Oil Pumping Excessive (Single-acting Compressor)	Operating Cycle Abnormally Long	Outlet Water Temperature Above Normal	Piston Ring, Piston, Cylinder Wear Excessive	Piston Rod or Packing Wear Excessive	Receiver Pressure Above Normal	Receiver Safety Valve Pops	Starts Too Often	Valve wear and breakage normal
Pulley or Flywheel Loose					●						●													
Receiver, Drain More Often																							●	
Receiver Too Small																							●	
Regulation Piping Clogged			●																					
Resonant Pulsation (Inlet or Discharge)												●	●	●	●									●
Rod Packing Leaks	●				●	●			●	●														
Rod Packing Too Tight						●																		
Rod Scored, Pitted, Worn																				●				
Rotation Wrong	●	●	●																					
Runs Too Little *(2)*								●																
Safety Valve Defective												●	●									●		
Safety Valve Leaks	●				●				●	●		●		●										
Safety Valve Set Too Low													●									●		
Speed Demands Exceed Rating																	●							
Speed Lower Than Rating									●	●														
Speed Too High	●	●			●						●				●			●						
Springs Broken																								●
System Demand Exceeds Rating	●				●				●	●		●			●		●							
System Leakage Excessive	●				●				●	●		●			●		●					●		
Tank Ringing Noise					●																			
Unloader Running Time Too Long *(1)*																●								
Unloader or Control Defective	●	●	●	●	●	●			●	●	●	●	●	●	●		●		●	●	●	●	●	●

Table 21–3 Continued.

THE CAUSES	Air Discharge Temperature Above Normal	Carbonaceous Deposits Abnormal	Compressor Fails to Start	Compressor Fails To Unload	Compressor Noisy or Knocks	Compressor Parts Overheat	Crankcase Oil Pressure Low	Crankcase Water Accumulation	Delivery Less Than Rated Capacity	Discharge Pressure Below Normal	Excessive Compressor Vibration	Intercooler Pressure Above Normal	Intercooler Pressure Below Normal	Intercooler Safety Valve Pops	Motor Over-heating	Oil Pumping Excessive (Single-acting Compressor)	Operating Cycle Abnormally Long	Outlet Water Temperature Above Normal	Piston Ring, Piston, Cylinder Wear Excessive	Piston Rod or Packing Wear Excessive	Receiver Pressure Above Normal	Receiver Safety Valve Pops	Starts Too Often	Valve wear and breakage normal
Unloader Parts Worn or Dirty				●																				
Unloader Setting Incorrect	●	●	●	●	●				●	●		●	●	●	●	●	●		●	●	●	●	●	
V-belt or Other Misalignment				●	●						●													●
Valves Dirty	●	●			●							●	●											
Valves Incorrectly Located	●	●		●	●				●	●		●H	●L	●H	●L	●			●H	●H				
Valves Not Seated in Cylinder	●	●		●	●				●	●		●H	●L	●H	●L	●			●H	●H				
Valves Worn or Broken	●	●		●	●				●	●		●H	●L	●H	●L			●H	●H	●H	●H			
Ventilation Poor	●	●			●										●									
Voltage Abnormally Low			●												●									
Water Inlet Temperature Too High	●	●			●				●			●						●						
Water Jacket or Cooler Dirty	●	●																						
Water Jackets or Intercooler Dirty						●						●						●						
Water Quantity Insufficient	●				●							●						●						
Wiring Incorrect			●																					
Worn Valve on Good Seat																								●
Wrong Oil Type		●																			●	●		
(1) Use Automatic Start/Stop Control																								
(2) Use Constant Speed Control																								
(3) Change to Non-detergent Oil																								
H (in High Pressure Cylinder)																								
L (in Low Pressure Cylinder)																								

compressor. In most cases, the dominant vibration frequency is the second harmonic (2×) of the main crankshaft's rotating speed. Again, this results from the impact that occurs when each piston changes directions (i.e., two impacts occur during one complete crankshaft rotation).

Valves. Valve failure is the dominant failure mode for reciprocating compressors. Because of their high cyclic rate, which exceeds 80 million cycles per year, inlet and discharge valves tend to work-harden and crack.

Lubrication systems. Poor maintenance of lubrication-system components, such as filters and strainers, typically causes premature failure. Such maintenance is crucial to reciprocating compressors because they rely on the lubrication system to provide a uniform oil film between closely fitting parts (e.g., piston rings and the cylinder wall). Partial or complete failure of the lube system results in catastrophic failure of the compressor.

Pulsation. Reciprocating compressors generate pulses of compressed air or gas that are discharged into the piping that transports the air or gas to its point(s) of use. This pulsation often generates resonance in the piping system, and pulse impact (i.e., standing waves) can severely damage other machinery connected to the compressed-air system. Although this behavior does not cause the compressor to fail, it must be prevented to protect other plant equipment. Note, however, that most compressed-air systems do not use pulsation dampers.

Each time the compressor discharges compressed air, the air tends to act like a compression spring. Because it rapidly expands to fill the discharge piping's available volume, the pulse of high-pressure air can cause serious damage. The pulsation wavelength, λ, from a compressor having a double-acting piston design can be determined by

$$\lambda = \frac{60a}{2n} = \frac{34,050}{n}$$

where λ = Wavelength, feet
        a = Speed of sound = 1,135 feet/second
        n = Compressor speed, revolutions/minute

For a double-acting piston design, a compressor running at 1,200 rpm will generate a standing wave of 28.4 feet. In other words, a shock load equivalent to the discharge pressure will be transmitted to any piping or machine connected to the discharge piping and located within twenty-eight feet of the compressor. Note that, for a single-acting cylinder, the wavelength will be twice as long.

Imbalance. Compressor inertial forces may have two effects on the operating dynamics of a reciprocating compressor, affecting its balance characteristics. The first is a force in the direction of the piston movement, which is displayed as impacts in a vibration profile as the piston reaches top and bottom dead center of its stroke. The second effect is a couple, or moment, caused by an offset between the axes of two or more pistons on a common crankshaft. The interrelationship and magnitude of these two effects depend upon such factors as (1) number of cranks; (2) longitudinal and angular arrangement; (3) cylinder arrangement; and (4) amount of counterbalancing possible. Two significant vibration periods result, the primary at the compressor's rotation speed ($\times$) and the secondary at 2$\times$.

Although the forces developed are sinusoidal, only the maximum (i.e., the amplitude) is considered in the analysis. Figure 21-1 shows relative values of the inertial forces for various compressor arrangements.

CRANK ARRANGEMENTS	FORCES		COUPLES	
	PRIMARY	SECONDARY	PRIMARY	SECONDARY
SINGLE CRANK	F' WITHOUT COUNTERWTS. 0.5F' WITH ● COUNTERWTS.	F''	NONE	NONE
TWO CRANKS AT 180° IN LINE CYLINDERS	ZERO	2F''	F'D WITHOUT COUNTERWTS. $\frac{F'D}{2}$ WITH COUNTERWTS.	NONE
OPPOSED CYLINDERS	ZERO	ZERO	NIL	NIL
TWO CRANKS AT 90°	1.41 F' WITHOUT COUNTERWTS. 0.707 F' WITH COUNTERWTS.	ZERO	707F'D WITHOUT COUNTERWTS. 0.354F'D WITH COUNTERWTS.	F'D
TWO CYLINDERS ON ONE CRANK CYLINDERS AT 90°	F' WITHOUT COUNTERWTS. ZERO WITH COUNTERWTS.	1.41 F''	NIL	NIL
TWO CYLINDERS ON ONE CRANK OPPOSED CYLINDERS	2F' WITHOUT COUNTERWTS. F' WITH COUNTERWTS.	ZERO	NONE	NIL
THREE CRANKS AT 120°	ZERO	ZERO	3.46F'D WITHOUT COUNTERWTS. 1.73F'D WITH COUNTERWTS.	3.46 F'D
FOUR CYLINDERS CRANKS AT 180°	ZERO	4F''	ZERO	ZERO
CRANKS AT 90°	ZERO	ZERO	1.41F'D WITHOUT COUNTERWTS. 0.707F'D WITH COUNTERWTS.	4.0F''D
SIX CYLINDERS	ZERO	ZERO	ZERO	ZERO

F' = PRIMARY INERTIA FORCE IN LBS.
$F' = .0000284\ RN^2W$
F'' = SECONDARY INERTIA FORCE IN LBS.
$F'' = \frac{R}{L}F'$
R = CRANK RADIUS, INCHES
N = R.P.M.
W = RECIPROCATING WEIGHT OF ONE CYLINDER, LBS
L = LENGTH OF CONNECTING ROD, INCHES
D = CYLINDER CENTER DISTANCE

Figure 21–1 Unbalanced inertial forces and couples for various reciprocating compressors.

STANDARD GRAPHICAL SYMBOLS

THE SYMBOLS SHOWN CONFORM TO THE AMERICAN NATIONAL STANDARDS INSTITUTE (ANSI) SPECIFICIATIONS. BASIC SYMBOLS CAN BE COMBINED IN ANY COMBINATION. NO ATTEMPT IS MADE TO SHOW ALL COMBINATIONS.	
LINES AND LINE FUNCTIONS	**PUMPS**
LINE, WORKING	
LINE, PILOT (L > 20W)	PUMP, SINGLE FIXED DISPLACEMENT
LINE, DRAIN (L < 5W)	
CONNECTOR	PUMP, SINGLE VARIABLE DISPLACEMENT
LINE, FLEXIBLE	**MOTORS AND CYLINDERS**
LINE, JOINING	MOTOR, ROTARY, FIXED DISPLACEMENT
LINE, PASSING	MOTOR, ROTARY VARIABLE DISPLACEMENT
DIRECTION OF FLOW, HYDRAULIC PNEUMATIC	MOTOR, OSCILLATING
LINE TO RESERVOIR ABOVE FLUID LEVEL BELOW FLUID LEVEL	CYLINDER, SINGLE ACTING
LINE TO VENTED MANIFOLD	CYLINDER, DOUBLE ACTING
PLUG OR PLUGGED CONNECTION	CYLINDER, DIFFERENTIAL ROD
RESTRICTION, FIXED	CYLINDER, DOUBLE END ROD
RESTRICTION, VARIABLE	CYLINDER, CUSHIONS BOTH ENDS

274

MISCELLANEOUS UNITS	
DIRECTION OF ROTATION (ARROW IN FRONT OF SHAFT)	
COMPONENT ENCLOSURE	
RESERVOIR, VENTED	
RESERVOIR, PRESSURIZED	
PRESSURE GAGE	
TEMPERATURE GAGE	
FLOW METER (FLOW RATE)	
ELECTRIC MOTOR	
ACCUMULATOR, SPRING LOADED	
ACCUMULATOR, GAS CHARGED	
FILTER OR STRAINER	
HEATER	
COOLER	
TEMPERATURE CONTROLLER	
INTENSIFIER	
PRESSURE SWITCH	
BASIC VALVE SYMBOLS	
CHECK VALVE	
MANUAL SHUT OFF VALVE	
BASIC VALVE ENVELOPE	
VALVE, SINGLE FLOW PATH, NORMALLY CLOSED	

BASIC VALVE SYMBOLS (CONT.)	
VALVE, SINGLE FLOW PATH, NORMALLY OPEN	
VALVE, MAXIMUM PRESSURE (RELIEF)	
BASIC VALVE SYMBOL, MULTIPLE FLOW PATHS	
FLOW PATHS BLOCKED IN CENTER POSITION	
MULTIPLE FLOW PATHS (ARROW SHOWS FLOW DIRECTION)	
VALVE EXAMPLES	
UNLOADING VALVE, INTERNAL DRAIN, REMOTELY OPERATED	
DECELERATION VALVE, NORMALLY OPEN	
SEQUENCE VALVE, DIRECTLY OPERATED, EXTERNALLY DRAINED	
PRESSURE REDUCING VALVE	
COUNTER BALANCE VALVE WITH INTEGRAL CHECK	
TEMPERATURE AND PRESSURE COMPENSATED FLOW CONTROL WITH INTEGRAL CHECK	
DIRECTIONAL VALVE, TWO POSITION, THREE CONNECTION	
DIRECTIONAL VALVE, THREE POSITION, FOUR CONNECTION	
VALVE, INFINITE POSITIONING (INDICATED BY HORIZONTAL BARS)	

METHODS OF OPERATION	
PRESSURE COMPENSATOR	
DETENT	
MANUAL	
MECHANICAL	
PEDAL OR TREADLE	
PUSH BUTTON	

METHODS OF OPERATION	
LEVER	
PILOT PRESSURE	
SOLENOID	
SOLENOID CONTROLLED, PILOT PRESSURE OPERATED	
SPRING	
SERVO	

GLOSSARY

Absolute	A measure having as its zero point or base the complete absence of the entity being measured.
Absolute pressure	The pressure above zero absolute, i.e., the sum of atmospheric and gauge pressures. In vacuum-related work, it is usually expressed in millimeters of mercury (mm Hg).
Accumulator	A container in which fluid is stored under pressure as a source of fluid power.
Actuator	A device for converting hydraulic energy into mechanical energy. A motor, cylinder, etc.
Aeration	Air in the hydraulic fluid. Excessive aeration causes the fluid to appear milky and components to operate erratically because of the compressibility of the air trapped in the fluid.
Annular area	A ring-shaped area. Often refers to the net effective area of the rod side of a cylinder piston. The piston area minus the cross-sectional area of the rod.
Atmosphere (one)	A pressure measure equal to 14.7 psi.
Atmospheric pressure	Pressure exerted by the atmosphere at any specific location.
Back pressure	A pressure in series. Usually refers to pressure existing on the discharge side of a load. It adds to the pressure required to move the load.

Baffle	A device or plate installed in a reservoir to separate the pump inlet from the return lines.
Bleed off	To divert a specific controllable portion of pump delivery directly to the reservoir.
Breather	A device that permits air to move in and out of a container or component to maintain atmospheric pressure.
Bypass	A secondary passage for fluid flow.
Cartridge	1. The replacement element of a fluid filter.
	2. The pumping unit from a vane pump, composed of the rotor, ring, vanes, and side plates.
Cavitation	A localized gaseous condition within a liquid stream that occurs where the pressure is reduced to vapor pressure.
Chamber	A compartment within a hydraulic unit. May contain elements to aid in operation or control of the unit.
Channel	A fluid passage, the length of which is large with respect to its cross-sectional dimension.
Charge	1. To replenish a hydraulic system above atmospheric pressure.
	2. To fill an accumulator with fluid under pressure.
Charge pressure	The pressure, above atmospheric, at which replenishing fluid is forced into the hydraulic system.
Check valve	A valve that permits flow of fluid in one direction only.
Choke	A restriction, the length of which is large with respect to its cross-sectional dimension.
Circuit	An arrangement of interconnected components to perform a specific function within a system.
Closed center valve	A valve where all ports are blocked in the center or neutral position.
Closed loop	A system in which the output of one or more elements is compared to some other signal to provide an actuating signal to control the output of the loop.
Compensator control	A displacement control for variable pumps and motors that alters displacement in response to pressure changes in the system as related to its adjusted pressure setting.
Component	A single pneumatic or hydraulic unit.

Compressibility	The change in volume of a unit volume of a fluid or gas when it is subjected to a unit change in pressure.
Control	A device used to regulate the function of a unit.
Cooler	A heat exchanger used to remove heat from the hydraulic fluid.
Counterbalance valve	A pressure control valve that maintains back pressure to prevent a load from falling.
Cracking pressure	The pressure at which a pressure-actuated valve begins to pass fluid.
Cushion	A device sometimes built into the ends of a hydraulic cylinder that restricts the flow of fluid at the outlet port, thereby arresting the motion of the piston rod.
Cylinder	A device that converts fluid power into linear mechanical force and motion.
Deadband	The region or band of no response where an error signal will not cause a corresponding actuation of the controlled variable.
Decompression	The slow release of confined fluid or gas to gradually reduce pressure on the fluid.
Delivery	The volume of fluid discharged by a pump in a give time. Usually expressed in gallons per minute (gpm).
Differential cylinder	Any cylinder in which the two opposed piston areas are not equal.
Directional valve	A valve that selectively directs or prevents fluid flow to desired channels.
Displacement	The quantity of fluid that can pass through a pump, motor, or cylinder in a single revolution, stroke, or cycle.
Double-acting cylinder	A cylinder in which fluid force can be applied to the movable element in both directions.
Drain	A passage in, or a line from, a hydraulic component that returns leakage fluid to the reservoir or to a vented manifold.
Efficiency	The ratio of output to input. Volumetric efficiency of a pump is the actual output in gpm divided by the theoretical or design output. The overall efficiency of a hydraulic system is the output power divided by the input power. Efficiency is usually expressed as a percent.
Enclosure	A rectangle drawn around a graphical component or components to indicate the limits of an assembly.

Energy	The ability or capacity to do work. Measured in units of work.
Feedback	The output signal from a feedback element.
Feedback loop	Any closed circuit consisting of one or more forward elements and one or more feedback elements.
Filter	A device whose primary function is the retention by a porous medium of insoluble contaminants from a fluid or gas.
Flooded	A condition where the pump inlet is charged by placing the reservoir oil level above the pump inlet port.
Flow control valve	A valve that controls the flow rate of oil.
Flow rate	The volume, mass, or weight of a fluid passing through any conductor per unit of time.
Fluid	1. A liquid or gas
	2. A liquid that is specifically compounded for use as a power-transmitting medium in a hydraulic system.
Force	Any push or pull measured in units of weight. In hydraulics, total force is expressed by the product of P (force per unit area) and the area of the surface on which the pressure acts: $F = P \times A$.
Four-way valve	A directional valve having four flow paths.
Frequency	The number of times an action occurs in a unit of time. Frequency is the basis of all sound. A pump or motor's basic frequency is equal to its speed in revolutions per second multiplied by the number of pumping chambers.
Full flow	In a filter, the condition where all the fluid must pass through the filter element or medium.
Gauge pressure	A pressure scale that ignores atmospheric pressure. Its zero point is 14.7 psi absolute.
Head	The height of a column or body of fluid above a given point expressed in linear units. Head is often used to indicate gauge pressure. Pressure is equal to the height times the density of the fluid.
Heat	The form of energy that has the capacity to create warmth or to increase the temperature of the substance. Any energy that is wasted or used to overcome friction is converted to heat. Heat is measured in calories or British thermal units (BTUs). One BTU is the amount of heat required to raise the temperature of 1 pound of water 1 degree Fahrenheit.

Heat exchanger	A device that transfers heat through a conducting wall from one fluid to another.
Horsepower (hp)	The power required to lift 550 pounds 1 foot in 1 second or 33,000 pounds 1 foot in 1 minute. A horsepower is equal to 746 watts or to 42.4 British thermal units per minute.
Hydraulic balance	A condition of equal opposed hydraulic forces acting on a part in a hydraulic component.
Hydraulic control	A control that is actuated by hydraulically induced forces.
Hydraulics	Engineering science pertaining to liquid pressure and flow.
Hydrodynamics	Engineering science pertaining to the energy of liquid flow and pressure.
Hydrostatics	Engineering science pertaining to the energy of liquids at rest.
Kinetic energy	Energy that a substance or body has by virtue of its mass (weight) and velocity.
Laminar flow	A condition where the fluid particles move in continuous parallel paths. Streamline flow.
Leverage	A gain in output force over input force by sacrificing the distance moved. Mechanical advantage or force multiplication.
Lift	The height a body or column of fluid is raised. Lift is sometimes used to express a negative pressure or vacuum. The opposite of head.
Line	A tube, pipe, or hose that acts as a conductor of hydraulic fluid.
Linear actuator	A device for converting hydraulic energy into linear motion, i.e., a cylinder or ram.
Manifold	A fluid conductor that provides multiple connection ports.
Manual control	A control actuated by the operator, regardless of the means of actuation.
Manual override	A means of manually actuating an automatically controlled device.
Maximum pressure valve	See Relief valve.
Mechanical control	Any control actuated by linkage, gears, screws, cams, or other mechanical elements.

Meter	To regulate the amount or rate of fluid flow.
Micron	One-millionth of a meter or about 0.00004 inch.
Micron rating	The size of the particles a filter will remove.
Motor	A device that converts hydraulic fluid power into mechanical force and motion. It usually provides rotary mechanical motion.
Open center circuit	One in which pump delivery flows freely through the system and back to the reservoir in the neutral position.
Open center valve	One in which all ports are interconnected and open to each other in the center or neutral position.
Orifice	A restriction, the length of which is small in respect to its cross-sectional dimension.
Passage	A machined or cored fluid conducting path that lies within or passes through a component.
Pilot pressure	An auxiliary pressure used to actuate or control hydraulic components.
Pilot valve	An auxiliary valve used to control the operation of another valve. The controlling stage of a two-stage valve.
Piston	A cylindrical part that fits within a cylinder and transmits or receives motion by means of a connecting rod.
Plunger	A cylindrical part that has only one diameter and is used to transmit thrust. A ram.
Poppet	That part of certain valves that prevents flow when it closes against a seat.
Port	An internal or external terminus of a passage in a component.
Positive displacement	A characteristic of a pump or motor that has the inlet positively sealed from the outlet so that fluid cannot recirculate in the component.
Power	Work per unit of time. Measured in horsepower (hp) or watts.
Power pack	An integral power supply unit usually containing a pump, reservoir, relief valve, and directional control.
Precharge pressure	The pressure of compressed gas in an accumulator prior to the admission of liquid.
Pressure	Force per unit area. Usually expressed in pounds per square inch (psi).

Pressure drop	The difference in pressure between any two points of a system or a component.
Pressure line	The line carrying the fluid from the pump outlet to the pressurized part of the actuator.
Pressure override	The difference between the cracking pressure of a valve and the pressure reached when the valve is passing full flow.
Pressure plate	A side plate in a vane pump or motor cartridge on the pressure port side.
Pressure-reducing valve	A valve that limits the maximum pressure at its outlet regardless of the inlet pressure.
Pressure switch	An electric switch operated by fluid pressures.
Proportional flow	In a filter, the condition where part of the flow passes through the filter element in proportion to pressure drop.
Pump	A device that converts mechanical force and motion into hydraulic fluid power.
Ram	A single-acting cylinder with a single-diameter plunger rather than a piston and rod. The plunger in a ram-type cylinder.
Reciprocation	Back-and-forth straight-line motion or oscillation.
Regenerative circuit	A piping arrangement for a differential-type cylinder in which discharge fluid from the rod end combines with pump delivery to be directed into the head end.
Relief valve	A pressure-operated valve that bypasses pump delivery to the reservoir, limiting system pressure to a predetermined maximum limit.
Replenish	To add fluid to maintain a full hydraulic system.
Reservoir	A container for storage of liquid in a fluid power system.
Restriction	A reduced cross-sectional area in a line or passage that produces a pressure drop.
Return line	A line used to carry exhaust fluid from the actuator back to sump.
Reversing valve	A four-way directional valve used to reverse a double-acting cylinder or reversible motor.
Rotary actuator	A device for converting hydraulic energy into rotary motion.
Sequence	1. The order of a series of operations or movements.
	2. To divert flow to accomplish a subsequent operation or movement.

Sequence valve	A pressure-operated valve that, at its setting, diverts flow to a secondary line while holding a predetermined minimum pressure in the primary line.
Servo valve	1. A valve that modulates output as a function of an input command.
	2. A follow valve.
Signal	A command or indication of a desired position or velocity.
Single-acting cylinder	A cylinder in which hydraulic energy can produce thrust or motion in only one direction.
Slip	Internal leakage of hydraulic fluid.
Spool	A term loosely applied to almost any moving cylindrical part of a hydraulic component that moves to direct flow through the component.
Strainer	A coarse filter.
Streamline flow	See Laminar flow.
Stroke	1. The length of travel of a piston or plunger.
	2. To change the displacement of a variable displacement pump or motor.
Subplate	An auxiliary mounting for a hydraulic component providing a means of connecting piping to the component.
Suction line	The hydraulic line connecting the pump inlet port to the reservoir or sump.
Sump	A reservoir.
Supercharge	See Charge.
Surge	A transient rise of pressure or flow.
Swash plate	A stationary canted plate in an axial-type piston pump that causes the pistons to reciprocate as the cylinder barrel rotates.
Tank	The reservoir or sump.
Throttle	To permit passing of a restricted flow. May control flow rate or create a deliberate pressure drop.
Torque	A rotary thrust. The turning effort of a fluid motor usually expressed in inch-pounds.
Torque converter	A rotary fluid coupling that is capable of multiplying torque.

Turbine	A rotary device that is actuated by the impact of a moving fluid against blades or vanes.
Turbulent flow	A condition where the fluid particles move in random paths rather than in continuous parallel paths.
Two-way valve	A directional control valve with two flow paths.
Unload	To release flow, usually directly to the reservoir, to prevent pressure being imposed on the system or portion of the system.
Unloading valve	A valve that bypasses flow to the tank when a set pressure is maintained on its pilot port.
Vacuum	Pressure less than atmospheric pressure. It is usually expressed in inches of mercury (in Hg).
Valve	A device that controls fluid flow direction, pressure, or flow rate.
Velocity	1. The speed of flow through a hydraulic line. Expressed in feet per second (fps) or inches per second (ips).
	2. The speed of a rotating component measured in revolutions per minute (rpm).
Vent	1. To permit opening of a pressure control valve by opening its pilot port.
	2. An air-breathing device on a fluid reservoir.
Viscosity	A measure of the internal friction or the resistance of a fluid to flow.
Viscosity index	A measure of the viscosity–temperature characteristics of a fluid as referred to that of two arbitrary reference fluids.
Volume	1. The size of a space or chamber in cubic units.
	2. Loosely applied to the output of a pump in gallons per minute (gpm).
Wobble plate	A rotating canted plate in an axial-type piston pump, which pushes the pistons into their bores as it "wobbles."
Work	Exerting a force through a definite distance. Work is measured in units of force multiplied by distance.

INDEX